THE ENERGY CONTROVERSY

THE FIGHT OVER NUCLEAR POWER

THE ENERGY CONTROVERSY

THE FIGHT OVER NUCLEAR POWER

FRED H. SCHMIDT • *DAVID BODANSKY*

UNIVERSITY OF WASHINGTON

ALBION PUBLISHING COMPANY
1736 Stockton Street
San Francisco, California 94133

Table of Contents

		Page
Foreword — Professor Hans A. Bethe, Cornell University		vii
Preface		xi

Chapter

1. The Energy Dilemma	1
2. What Is Energy?	5
3. The Supply of Fossil Fuels	11
4. Renewable Resources	21
5. The Potential of Nuclear Energy	27
6. Residues From Energy Production	37
7. Characteristics of Nuclear Energy	41
8. Comparative Costs	45
9. Why the Dispute?	51
10. Radiation Hazards	55
11. Plutonium and Human Health	59
12. Handling the Wastes	65
13. Are Nuclear Reactors Safe?	73
14. Breeder Reactors—Prospects and Problems	93
15. Safeguards and Security	101
16. Nuclear Proliferation	111
17. Moratorium Madness	115
18. The Need for Nuclear Energy	121

Appendices

A. Energy Conversion Chart	129
B. Calculation of Increase of Temperature of the Earth	131
C. The Linearity Assumption for Radiation Effects	133
D. Test of the Emergency Core Cooling System	139
E. Nuclear Liability Insurance and the Price-Anderson Act	143
Bibliography	149

FOREWORD

by Hans A. Bethe, Nobel Laureate, Cornell University

For many Americans, the Energy Crisis is a thing of the past: It happened in early 1974 and it consisted of long lines at gasoline stations; it was over when the Arabs lifted their oil embargo. It was an impressive, a bit frightening, but short experience, "probably manipulated by the oil industry." After all, you can now go to a gas station and buy all the gasoline you want and your heating oil is delivered regularly. Perhaps you keep your thermostat 2 degrees lower, but the only real reminder is higher prices. And if you live in the Northeast and build a new home, the local utility very likely will be unable to supply you with natural gas.

In reality, the energy problem is only beginning. The oil embargo was a useful signal that times are changing; oil is running out. The best estimates of the United States Geological Survey and the National Academy of Sciences tell us that without imports, the total domestic resources of oil would supply our present consumption for only 25 more years. The proven reserves are much less. Mideast oil will last longer; if the demand continues at present levels, it may last 50 years. But this does not allow for any increase in demand, either from developing or industrial countries.

So the industrial world is facing a transition from fluid fuels (oil and natural gas) to something else. This transition creates what is probably the most severe problem since the beginning of the industrial age. Industrial development has taken two centuries, but the conversion to new energy sources will have to be accomplished in about two decades. Speed is essential, particularly because the construction of any major energy facility takes from five to ten years.

The solution of this problem is also essential from the political point of view. Only if we regain a considerable degree of energy independence can we successfully resist political pressure from the OPEC (Organization of Petroleum Exporting Countries), some of which nations have political aims opposed to our own.

The problem could be solved if we had strong leadership and a generally recognized national purpose, such as we had in World War II, and rational discussions of the best measures to pursue. This is not the case. Instead, we have angry recriminations on many of the issues involved, a debate which does not clarify anything.

When I am asked for my solution to the energy problem, I say "coal and nuclear energy."

"Nuclear meaning fusion?" comes back the second question.

"No, fission," is my reply. "This is the only nuclear energy we know how to use."

"But," continues the questioner, "fission is full of dangers."

Then I must patiently explain why fission is not as full of dangers as the questioner assumes. It is of necessity, a long explanation, and about halfway through, the questioner's attention begins to wander and his fears hold the upper hand.

From now on, I can refer my questioners to this book by Fred Schmidt and David Bodansky. It gives the answers to the puzzled and curious layman who really wants to become informed.

The book discusses nuclear power in a rational and unemotional way. It explains the important technical problems and does not avoid the objections raised by the opposition. The style is lucid and simple, which should make it easy reading for any educated layman. Similes from everyday life are brought in to make the concepts clearer. The book is short enough to be easily readable yet long enough to cover the subject.

The authors freely acknowledge there are problems in nuclear power. But they also show that most of them can be solved. While they do not claim that nuclear energy is perfect, they prove that it is reasonably safe.

Neither of the authors is or has been employed by the nuclear industry or by government agencies responsible for nuclear power. Their livelihood does not depend upon nuclear power. Both are scientists. They have studied diligently the design of nuclear reactors, their performance, and their problems. They have come to the conviction (as I have) that nuclear power is acceptable from the standpoint of safety, of medical effects, and of cost, and that it is needed for the well-being of the nation and the economy. They want to make an understanding of the problems easier for the reader, and save him the many months of study they themselves have devoted to the issue.

Is nuclear energy (aside from coal) really our only choice? Everyone has his pet solution of the energy problem (if he admits there is a problem at all)— solar energy in its various forms, windmills, tides and ocean waves, biomass, geothermal energy, nuclear fusion, to name just a few. All of these, or nearly all, should be subjects of intensive research, and, indeed, ERDA (the United States Energy Research and Development Administration) is sponsoring intensive research on most of them. However, it is essential to distinguish between research that can bring new energy sources into production in ten to twenty-five years, and longterm sources. If research to establish a method for using a given energy source is still proceeding, this method, can in general, become available only in the long term.

It takes a very long time from the birth of an idea to the point of proving its value in the laboratory, and a much longer time for engineering development to make the process usable in a large industrial plant. And it takes a still longer time before a major industry can be established. Certainly for the next 10 years and probably for the next 25 years the United States can not expect any of the proposed alternative energy schemes to have much impact.

Thus, we must use what is now available—fission and coal. Perhaps the nuclear power age will last only 50 years; perhaps we will find another solution by then which is no more expensive and is publicly more acceptable. Perhaps . . . but it would be foolhardy to rely on it.

Therefore, we must do the best we can with nuclear energy. First of all, we must keep it safe and make it still safer through appropriate research. We must have a sensible policy on the siting of nuclear power plants, keeping them away from population centers, but at the same time making it possible to find *some* location in those areas in which power is needed. We must develop the best safeguards against misuse of nuclear materials for weapons and terrorist activities, while not interfering with the civil liberties of the population at large. All this I believe can be done. Finally, we must regulate international trade in such a way that nuclear power is available to the countries that need it, but not nuclear weapons; this is the most difficult task.

The book by Schmidt and Bodansky fills a great need. It supplies public information in a field of greatest importance—information which has been obscured by the heat of a public debate carried on at a very low level. The public should get its information from fact, not fiction.

Hans A. Bethe

PREFACE

Pre-prints of this book were first distributed in February, 1975, to many physicists and other scientists, to engineers, and to environmentalists, as well as to other persons with diverse backgrounds and interests. We received gratifying responses from many of them, as well as suggestions and criticisms, all of which were given considerable thought. This resulted in a new version distributed in August, 1975, and ultimately in the present book, which represents an expansion and revision of the August text.

Since our original writing, many new studies and reports on the subject have appeared. Accordingly, the book was revised to take into account the new material and the suggestions and criticisms we have received. In addition, we have included substantial newly developed information, particularly in relation to breeder reactors, plutonium hazards, reactor safety, and waste disposal. Indeed, the intensity of the nuclear debate has grown so that we have changed the title from *The Role of Nuclear Power* to *The Fight Over Nuclear Power*.

Many friends, colleagues, and physicists with whom we are acquainted have asked what we are "doing" about the energy problem and how we happened to get involved. In research, experimentalists—such as we are—conduct experiments which hopefully will teach us something new about nature; theoreticians develop and consider new theories and try to explain existing facts or predict new experimental facts. In our energy studies, we have done none of these things. Instead, we have returned to the three R's; we read, we do arithmetic, and we write about it, in that order.

Our experience as physicists helps us reduce the vast array of numbers describing energy sources, future demand scenerios, etc., to what we hope are easily understood and significant facts. We think our experience also has helped us seek out the *most important* features of the controversy and avoid becoming bogged down in interesting but unimportant side issues.

In writing, we hope our experience in teaching the elements of physics to non-physics majors has helped us communicate with non-scientific readers perusing this energy study. Nevertheless, the energy problem is *not* simple, and we can only hope that we have been reasonably successful in this effort.

That leaves our friends' second question: How did we get into it?

Early in 1973, we first became interested in energy issues, and what we learned, especially about the limitations of oil resources, was disturbing. That summer and fall, we organized formal and informal seminars, and we soon found ourselves attending other seminars, joining symposia, and traveling to American Physical Society and other professional meetings in the United States and Europe to hear energy experts examine and debate different aspects of the problem.

Although we are nuclear physicists, we decided in the beginning that we would not concentrate on nuclear energy, because we were not convinced it held the best promise of a solution. Also, we did not wish to follow what might have been an easy path for nuclear scientists. As may be evident to our readers, our early opinions were altered in the course of the study; if the world held a gigantic reservoir of natural gas, our conclusions might be different.

It is perhaps relevant to comment on what being a nuclear physicist means. In the Second World War, virtually all nuclear physicists (and there were very few by today's count) worked on the Manhattan Atomic Bomb Project. After the War, most nuclear physicists returned to the pursuit of "fundamental" research—for example, studies of the structure of nuclei or of "elementary particles." They left the practical aspects of nuclear energy almost entirely to the newborn field of nuclear engineering, although many of the more fundamental studies found application in nuclear engineering.

Traditionally, that is exactly what happened in many other fields of research in physics, and it was quite proper that it should happen in this case as well. One of us proved to be no exception to the general pattern and shifted into fundamental research at war's end; the other entered the scene a bit later. As a result, we were scarcely better informed at first about nuclear power than most laymen.

Thus, it can be seen that, contrary to common opinion, which considers nuclear physicists and nuclear engineers as being embraced by a single discipline, a wide gulf separates the two. It is unfortunate that such a gulf exists, but for us at least, it has been partly bridged through this study. It is fair to say, however, that our knowledge of nuclear physics was helpful in enabling us to grasp more quickly the essential problems of nuclear power. We also believe our engineering experience aided our judgments of technical considerations.

More than half the fundamental nuclear research, as distinct from nuclear engineering, of the past quarter century has been assisted financially by the U.S. Atomic Energy Commission (AEC), through contracts and grants to universities. Our laboratory at the University of Washington was partly supported in this manner. The AEC never told us what to do nor how to do what we did. Our primary concern always was and remains the study of nuclear physics as we believe appropriate.

It was *our* choice to study the energy controversy and the expenses of our work have been borne by both our university and the university's contract with the AEC, and more recently, with the U.S. Energy Research and Development Administration, which took over many of the AEC's functions in the 1974 reorganization.

In the course of our studies, we have developed a degree of knowledge of energy issues, and that, quite naturally, has led us to a viewpoint and conclusions. The fact that we are immersed in the issue and have formed a definite viewpoint has of itself generated a suspicion among some persons that we have a vested interest and are thus not to be trusted. We find this attitude disturbing. How can anyone, or the nation as a whole, come to rational decisions in such a climate?

Our objective is to follow a logical sequence in considering the energy problem, albeit our logic, which, we hope, will lead to an "obvious" solution. We are not so naive as to think that an irresistibly logical case can be made, but we do believe that such efforts can contribute to the understanding of the controversy. Our conclusion is that nuclear fission energy is the most logical choice, for it is impressively safe, is to be preferred over fossil fuels, and indeed, may prove to be man's only alternative, not only for the next few decades, but for many years in the future.

We have included technical discussions to the extent which we believe necessary for an understanding of the issues and problems. Consequently, some readers may want to skip over sections with which they are already familiar, or which to them seem unnecessarily detailed.

In view of the dimensions of the problem and the diversity of opinions, we have included a large number and wide variety of specific references as an aid to the critical reader. A more general reading list is also appended. Some details, not essential to the mainstream of our arguments, are relegated to Appendices.

We wish to thank our friends, colleagues, and critics who read drafts of this document and to whom we are indebted for helpful suggestions. In particular, we thank Professor Hans A. Bethe, Physics Department, Cornell University; Professor Bernard L. Cohen, Physics Department, University of Pittsburgh; Dr. John Manley, former Associate Technical Director, Los Alamos Scientific Laboratory; Dr. J. Carson Mark, former head of the Theoretical Division, Los Alamos Scientific Laboratory; Professor Ove Nathan, Niels Bohr Institute, Copenhagen; Dr. William E. Siri, Lawrence Berkeley Laboratory and past President, Sierra Club; Professor Maurice A. Robkin and Gene L. Woodruff, Nuclear Engineering Department, University of Washington, and Mr. Robert W. Holloran, P.E., Seattle, for their specially valuable comments and assistance.

THE ENERGY DILEMMA
Chapter One

If balanced judgment and good sense do not prevail in the handling of America's dilemma over the production and use of energy, the nation could face a political and sociological crisis worse than that created by the Vietnam War. Never has there been a greater need for a willingness to avoid quick emotional reactions, and for a determination that decision-making be based on a sound analysis of the situation.

The dilemma has arisen because our energy shortages threaten to become progressively worse unless prompt and prudent action is taken. Future energy crises could make those of the past look like trivial occurrences.

America's energy problems first became widely recognized in the winter of 1972–73. They manifested themselves in the form of regional electric power or heating-oil shortages, and soon, shortages of natural gas. By the summer of 1973, gasoline shortages popped up occasionally here and there. Then, precipitated by the October War in the Middle East, the crisis reached its peak in the autumn and winter months of 1973–74, with severe shortages of oil, long waiting lines, and short tempers—to say nothing of skyrocketing prices and a bulging inflation.

Today, the immediate crisis has receded, but the shock remains from the discovery of America's dependence on imported oil. The causes of the shortages are still hotly debated, with blame fixed alternately on industry, the government, or the oil exporting countries. However, there is increasing recognition of deeper reasons for our energy problems, rooted in the disparity between demand for energy in the U.S. and the world, and the available resources. The crisis, or problem, or whatever nomenclature it bears, will not go away; searches for solutions have provoked national and international controversy.

1

The controversy has innumerable facets which complicate the steps that should be taken to make decisions designed to solve the dilemma. The problems include:

- The adequacy of the supply of fossil fuels.
- The pollution of the ocean shores and oceans by oil spills from supertankers.
- The contamination of the atmosphere by sulphur dioxide and particulates from coal burning.
- The dangers of natural gas explosions.
- The long-range effects of carbon dioxide in the atmosphere.
- The practicality of "new" energy sources, such as geothermal, solar, fusion, wind and wave power.
- The extent of environmental damage from strip coal mining.

The list could be extended, but the most dramatic problem, the one that stands out above all others in the controversy, is the nuclear dilemma. It has come to the fore because of the wide difference between the fears of the opponents and the claims of the proponents, and because nowhere else are the stakes so high and so crucial to the U.S. and the world.

The Nuclear Age was born under the mushroom cloud with the horror of the bomb. The bang was great and so was the hush of secrecy which perforce surrounded the military aspects—a secrecy that many believe spilled over into peaceful, practical applications. As the peaceful applications of nuclear power grew under the impetus of the AEC and the newborn private nuclear industry, communications with the public concerning the facts about nuclear energy failed. Simultaneously, however, nuclear engineering became a major discipline in most of our larger universities. Thus, the facts were laid open to those able to undertake detailed studies.

In the course of these developments, the problems escaped close public scrutiny. Neither the benefits nor the difficulties of nuclear power were widely aired. Even the extraordinary contributions of nuclear science to medical practice have remained little known or have been accepted without criticism or appreciation.

Therefore, it is not surprising that an aura of mystery has grown up around nuclear power, the more so since the concepts and technology of nuclear power are new and complex.

People are most often afraid of what they know little or nothing about; the frequently publicized dangers of radiation make it easy for some people to convince others that nuclear power is a threat to society.

Consider the enormous contrast between the pro and con viewpoints. Nuclear energy is widely attacked by environmental groups; yet a strong case can be made for stating that nuclear power is almost uniquely benign in its environmental implications. It is widely attacked on issues of safety; yet its proponents point to a record which demonstrates impressively that it is setting and maintaining unprecedentedly high standards for industrial safety—higher in fact, than for any comparable industry. It is attacked on grounds of prudence; yet the elimination of nuclear power is thought by many to be a gamble with extraordinary inherent peril.

Nuclear power is not alone in being surrounded by widespread misunderstandings. It has not been a pleasant decade for science and technology as a whole. Perhaps the practitioners are primarily to blame for the predicament. They have been remiss in devoting sufficient time and effort to the serious business of communicating what they are about to the non-scientists and non-engineers who make up the mass of society. If a news commentator, editor, magazine writer, politician, or citizen activist levels a tirade against some form of scientific endeavor or engineering process, the fault may be ascribed in part to science and technology for failing to convey intentions and methods in the first place.

The problem of public understanding is complicated further by the fact that unanimity among scientists and engineers is rare, even when broad consensus exists. In addition, the voices of dissenting "experts" are often widely publicized, particularly if they are voices of alarm.

One of the purposes of this book is to analyze the contradictory views of nuclear energy and to put the issues in some perspective for the public, which in America will make the ultimate decisions. Much of the difficulty in addressing the problem derives from the emotions evoked. On one hand is the fear engendered by anti-nuclear critics who focus on the dangers of radiation, and on the other the exasperation voiced by technically oriented proponents, who believe there is a limit to the deference which should be given to "unreasonable fears."

Perhaps it will never be possible to eradicate the connection between nuclear weapons and nuclear energy; indeed there is a close tie. The same "power of the atom" that erupted in those horrendous clouds over Hiroshima and Nagasaki is responsible for the nuclear-powered electricity now lighting and heating millions of homes in America and abroad.

But, there is a crucial difference. For nuclear weapons, every effort is made to enhance the destructive effects; for nuclear power plants, every effort is made to insure safety. A question remains: Have these efforts to provide safety been sufficient or are the problems too intractable to yield even to the best engineering practice and sociological effort?

It is our objective to follow a logical sequence in considering this question and the others which comprise the energy problem. We hope the process will lead to an "obvious" solution. We are not so naive as to believe an irresistible case can be made, but we believe such an effort can contribute to the reader's knowledge and understanding of the controversy.

Our study of the available fossil fuel and nuclear resources, when combined with judgments of the technical feasibilities of untried alternatives, leads us to conclude that nuclear fission energy is our most secure choice for the foreseeable future.

Coal resources may carry us for a time. In fact, we must rely in part, upon coal for the near-term. But we can not afford to delay in deploying nuclear power—power which is impressively safe, which is economically and environmentally preferable to coal, and which utilizes a resource having essentially no other use.

The hopes held by many that the world will soon have unlimited energy from the sun, the inner earth, fusion, and other sources are premature. Unfortunately, these hopes are not supported by current scientific and engineering evidence.

The human stakes involved in the decisions we make are very great, and they call for our best and most rational consideration. It is important that citizens everywhere carefully weigh the evidence concerning nuclear power. The time to act responsibly, sensibly, and logically is now—not a decade or two hence, when the hazards of emotionalism may have out-damaged any hazards of controlled radiation.

WHAT IS ENERGY?
Chapter Two

The concept of energy was invented by scientists a long time ago. It is usually thought of as something which produces or changes the motion of objects or which, equivalently, can produce heat.

A lump of coal, a cube of sugar, a barrel of oil, a kilogram of uranium are all examples of objects containing energy.

A truck hurtling down a freeway is another example. It has energy because it can collide with an automobile and thus impart motion to the automobile. The cube of sugar has energy because suitable "conversion devices"—for example, people—exist which can convert the energy of the sugar cube into motion.

A person could push an automobile with a force of, say, 100 pounds and move it, say, 10 feet, thus performing 1,000 foot-pounds of "work." In the process, he has expended 1,000 foot-pounds of sugar-cube energy he drew from his morning coffee.

Many other measures or units of energy exist like the foot-pound example. They include the kilowatt-hour and the food unit—the kilo-calorie or simply the calorie. Each has its own conversion factors.

Appendix A provides a chart of some of these factors. For example, one kilowatt-hour is equivalent to 860 kilo-calories. The reason a unit of time—the hour—appears in the "kilowatt-hour" term is that a kilowatt (a thousand watts) is actually a unit of power, which is the time-rate of expenditure of energy.

All these units of energy have come to be used because of convenience in a particular circumstance. Other examples are a ton of coal, a barrel of crude oil, a gallon of gasoline, a kilogram of uranium. All have energy equivalents and thus can be conveyed as energy units.

5

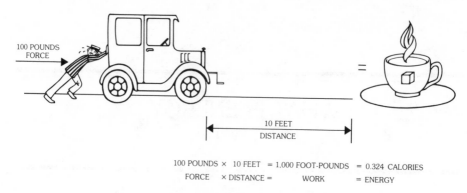

$$100 \text{ POUNDS} \times 10 \text{ FEET} = 1{,}000 \text{ FOOT-POUNDS} = 0.324 \text{ CALORIES}$$
$$\text{FORCE} \times \text{DISTANCE} = \text{WORK} = \text{ENERGY}$$

Man pushing automobile. He does "work", and thus expends energy. We convert foot-pounds to Calories by application of Appendix A.

The reference to man as a "conversion device" is significant. He is in effect a machine or engine which can use certain sources of energy for conversion into useful work. A source of energy becomes useful only if a suitable conversion device exists.

Nature has many such conversion devices. The sun is a spectacular example as it converts hydrogen into helium, thus releasing nuclear energy in the form of radiation.

Man has also invented some conversion devices. A match can start a chemical chain reaction in a pile of wood, which burns and produces heat. In turn, the heat can be used in an engine to produce motion. Most conversion devices are "heat engines," as in cars or in most electric-power plants.

An important characteristic of energy conversion devices is that they are generally not perfect. Each has a certain inefficiency. As a result, only a fraction, always less than 100 percent, of the initial energy is converted.

A person is about 25 percent efficient when he works hard, as he would in running, for example. The limits to the efficiency of conversion devices are imposed by physical laws and by practical considerations. To illustrate, it is evident technology has little chance of raising the efficiency of modern steam turbines much beyond 40 percent. That means 60 percent of the energy consumed in, say, a coal-fired electric-power plant is wasted—and it is usually wasted by heating up the atmosphere, a river, or the sea. (Increasingly, however, attention is being given to utilizing waste heat in industrial or domestic heating.)

It is important to note that *all* energy consumed by a power plant returns to the environment as heat. If we use the electricity to operate a television set, the heat produced warms our house and escapes to the atmosphere. Even a refrigerator produces heat which warms the kitchen.

Perhaps the most remarkable and most easily understood energy scale is the human body and its need for food-energy. An average person requires about 75 watts of power continuously simply to stay alive. That is the metabolic rate, which in more familiar units is one calorie per kilogram of body weight per hour. (A kilogram is equivalent to 2.2 pounds.)

On a daily basis the body requires 75 watts times 24 hours, or, roughly, 2 kilowatt-hours of energy a day. If that energy need were to be supplied by electricity in the U.S., the cost would be between 2 and 14 cents, depending upon the area in which one lives.

It should be obvious that it is not very useful for a person to operate at the very low level of activity corresponding to the base load. The 75-watt rate should be increased by another 50 or 60 watts to describe the average American's actual activity. Thus, the typical daily food-energy consumption is slightly more than 3 kilowatt-hours, or the equivalent of about 3,000 calories.

As a comparison, it is noted that the total per capita power consumption in the U.S. is nearly 12 kilowatts, which is about 300 kilowatt-hours a day or about 100 times the average human biological need. Even in 1850, when 90 percent of our energy was derived from the burning of wood, we used over 3 kilowatts per person.

The human body's daily food requirements cost far more than the 3 to 21 cents that would be required if the body ran on electricity. One reason is that for each unit of food-energy produced in the U.S. system of production, about 10 units of other kinds of energy are utilized.[1] The other kinds are consumed by the tractors, the trucks, the trains, the fertilizer plants, the irrigation pumps, and so on, that are involved in the intricate food-production chain.

With the inclusion of this multiplication factor, it can be seen that each person's daily food-energy equivalent amounts to about 35 kilowatt-hours. That is approximately 13 percent of all energy consumption in the U.S.[1,2] The situation is similar in other industrialized nations.

Therefore, it is apparent that food-energy is about the most expensive kind. For all that, it is much cheaper in an industrialized society than in a rural nation. As a measure of its relative price, it should be pointed out that food expenditures represent about 20 percent of the gross national product in the U.S.[3] whereas they are up to about 80 percent in underdeveloped countries.

Where is this enormous energy consumption utilized in the food system? A substantial portion of it goes into helping solar energy, as it were, produce more food than Nature could by herself.[4] This helping hand is one reason U.S. agriculture is so productive. As a rough estimate, American grain production would drop by a factor of two or more—less than half the present output—without this help or stimulus. The impact on *our* lives would be drastic enough, but it would be much greater on other regions of the world because we export some 20 percent of our grain.[2]

Another massive effect of cutting back the use of energy in producing food would be the impact on the American social order. Let us presume that such a cutback is mandated; that would force millions to return to the soil to grow their own food. It would be logical to question whether the U.S. has gone too far in the displacement of people from the farms; only one in 50 is a farmer today. However, the displacement was by choice, not order, and the vast demographic population shift from the farm to the city took place slowly in America, spread over many decades. Any rapid turnabout and a large-scale migration back to the farms would bring tremendous strains to the American society.

Viewed in the broader sense, the only hope for much of the rest of the world to acquire adequate food-energy is to emulate the U.S. agricultural system as far as possible or to reduce populations drastically—or both. By any conceivable humanitarian methods, the time-scale required to achieve sufficient population reduction would be far greater than that necessary to satisfy their needs by the energy-intensive route.

Although food production is the area in which energy needs are the most urgent, energy plays an important part in enhancing other aspects of the quality of life. About 20 percent of our energy consumption serves to heat our homes and places of work. The rest, 67 percent, is used in industry, transportation, and the removal of drudgery from chores at home or at work.

The result attained by the U.S. and other industrialized nations has been such that the rest of the world has signified it would like to enjoy similar status. Indeed, the percentage increase in world energy consumption each year is greater than for the U.S. alone. As a specific index of the effect of energy use on a society, it is noteworthy that energy consumption per capita worldwide is proportionate approximately to per capita income.[5] Thus, it can be seen that the American standard of living—from this indicator at least—is directly related to the amount of energy produced and utilized.

We arrive, therefore, at our first conclusion:

- In addition to taking all reasonable conservation measures, the U.S. and all other nations should adopt energy-production policies which will permit, as far as possible, an increased worldwide per capita energy supply.

The U.S. rate of increase can be slowed, but it is imperative that America assist the developing countries in achieving higher levels of energy consumption per person. Energy use can be cut back in the U.S., at least among the more prosperous, through more intelligent methods of use. For example, conservation steps might include improved thermal insulation of homes to reduce heat-loss, or a shift from grain-fed to grass-fed beef.

"Energy conservation" has a scientific meaning, as well. Energy is never really lost, it simply changes from one form to another, as, for example, in the conversion of a lump of coal to a roomful of warm air. It might be assumed, therefore, from this Law of Conservation of Energy, as it is called in physics, that we need only be clever enough to recapture the expended energy and use it over and over again, since it has not been destroyed.

Another law of physics, however, precludes this possibility. It's called the Second Law of Thermodynamics, and it informs us that once our lump of coal has been converted to heat at a low temperature, as in a reservoir like the ocean, we have lost its energy for good. Moreover, it must be remembered that devices converting heat to motion, like automobile engines, are limited in efficiency.

Incidentally, automobile engines have a low efficiency—20 percent at best—a fact which raises a fascinating point. We could conserve energy by burning gasoline to make electricity (at an efficiency of about 40 percent!) and running our automobiles on the electricity instead, because electric motors are very efficient! Unfortunately, we have not yet learned how to accomplish it on a large scale. It has been done on a relatively small scale with electric busses, but that application is limited. Also, battery-powered automobiles will work, but they are presently uneconomical because of the large capital investment.

All the energy constantly being poured into the environment constitutes a source of total energy pollution, and one could worry with some logic that the earth will heat up significantly as a result. Fortunately, another law of physics, a Law of Radiation, reduces the effect considerably. A very small increase in temperature causes a body such as the earth to lose energy to space at a greatly increased rate.

If the population of the world were tripled to 12 billion people, and everyone enjoyed using energy equal to the present U.S. rate per capita (about 10,000 watts), the increase in the earth's temperature would be about one-tenth of a degree Fahrenheit.[6] (The calculation is made in Appendix B.) That increase would be quite negligible when compared with the normal fluctuations of more than one degree in global temperature.[7]

Another way to recognize why the earth's temperature would not be "disturbed" very much is to compare the 12 billion times 10,000 watts with the energy received by the earth from the sun. It is only one tenth of a percent as much.

To be useful, an energy source must first pass muster in those two respects—that is, be plentiful and be technologically convertible. Beyond them, other tests must be applied—such tests as environmental impacts, health and safety requirements, economic considerations, and many others. With all these efforts and inputs, we should be able to choose, in principle, the best logical course of action at a given point in time—in particular, the present time.

In practice, unfortunately, such a logical approach works only imperfectly. We cannot make perfect technological assessments, and different people make different assessments. In the realm of political and social considerations, we find even greater disagreements. Nevertheless, it is our purpose to try to apply such a logical approach, and that is the subject matter of the chapters that follow.

One additional word is necessary regarding technological assessments. It is always possible that a technological "breakthrough" could be made which would substantially alter the options available. But the success of radically new techniques should not be relied upon *in advance* of their demonstration. The stakes for mankind are too great to run the risk of depending on major new discoveries, because they may not occur until some indefinite future time—which may be far beyond the time they are desperately needed.

REFERENCES

1. Carol and John Steinhart, *Energy Sources, Use and Role in Human Affairs,* p. 80. Duxbury Press, 1974.

2. J. Steinhart and C. Steinhart, *Energy Use in the U.S. Food System,* Science *184*, 307 (Apr. 1974).

3. Eric Hirst, *Food-Related Energy Requirements,* Science *184*, 134 (Apr. 1974); N. Thimmesch, Los Angeles Times (Seattle Times, March 19, 1974), quotes about 16 percent.

4. David Pimentel, et al, *Food Production and the Energy Crisis,* Science *182*, 443 (Nov. 1973).

5. D. H. Meadows, et al, *The Limits to Growth,* Potomac Associates (1972), p. 77.

6. James Tuck, Univ. of Wash. Public Lecture, Apr. 11, 1974.

7. Time Magazine, p. 80, Nov. 11, 1974.

THE SUPPLY OF FOSSIL FUELS
Chapter Three

Our chief sources of energy are coal, oil, natural gas, and gas liquids. They are referred to as fossil fuels because they were produced by biological conversion of solar energy long ago and over a great length of time. On time scales consistent with mankind's reasonable future existence, we must look upon these energy sources as nonrenewable.

Coal was the first to be discovered and was originally utilized by people simply as a source of heat. Today we use all three fossil fuels for that purpose. The devices for the conversion of fossil fuel to heat are conceptually rather simple, ranging from a Franklin stove or a household oil burner to the sophisticated coal-burning furnace of a modern electrical power plant.

Conversion of heat energy to useful motion by means of heat engines was accomplished relatively recently in the history of man, and its development brought on the industrial revolution. Subsequently, converting heat to electrical energy contributed vast new horizons by providing a very convenient system for energy transmission and for communications.

Coal continued as the chief source of energy in the U.S. until a comparatively short time ago. Beginning in the 1920s, oil began to supplant coal. Oil is more convenient for many uses, so its increased utilization came quite naturally, particularly since oil was found to exist in many places. It has been pointed out by the geologist, M. King Hubbert,[1] that between 1956 and 1968, a brief span of 12 years, the world consumption of oil equaled the amount consumed in all of history up to 1956! That startling fact alone should be sufficient to alert us and to stimulate us to a most careful examination of total geological reserves.

Natural gas is a newer source of energy than oil. Again, it is not surprising that natural gas gained in popularity; it can be transported by pipeline cheaply, the furnaces burning it are even simpler than oil-burners, and it burns with very little residue. Since natural gas and oil deposits are associated

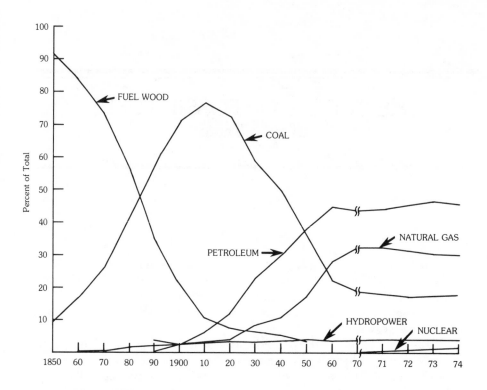

U.S. energy consumption patterns, 1850-1974. From *Energy Perspectives,* U.S. Department of the Interior, 1975.

geologically (with energy contents of about two for oil to one for gas), an examination of the U.S. and world reserves of oil will suffice for both fuels.

Our changing patterns of fuel usage are shown pictorially in the diagram. Today, oil and natural gas together account for about 75 percent of our total energy use.

When faced with the question of determining how plentiful an energy source is, we encounter the problem of making measurements and estimates for a substance hundreds to thousands of feet below the earth's surface or below the bottom of the sea. In addition, we must then properly interpret the significance of the vast array of numbers emerging from the estimates.

The first problem we encounter is the initially baffling array of terms. What is an "undiscovered, recoverable" resource? What is a "reserve"? If something is undiscovered, how do we know it is there? Or do we?

Fortunately, the terminology has now become fairly standard.[2] Resources are the total amount of, say, oil which can be economically extracted from the part of the earth one is considering. Thus one speaks of resources for Alaska, or for the continental U.S. Reserves are the part of the resources which have already been found, and whose locations and amounts are known. Undiscovered recoverable resources are the further amounts which are estimated to exist in usable form. These estimates are based on past experience and "broad geologic knowledge and theory." As can be imagined, there is much more disagreement among experts about the undiscovered recoverable resources than about the reserves. Even for the reserves there are distinctions between "demonstrated" reserves and "inferred" reserves, but we can here lump them together without further worry.

After we determine the extent of the resource, we must compare it to the expected or actual demand for that resource. Many efforts have been made to produce careful estimates of the future demand for oil and other energy sources. The predictions are based on various assumptions, such as the continuation of trends in oil consumption, expected population increases, and so forth. Detailed estimates of this nature are extremely useful to the specialists—an oil company, for example, whose directors must make hard and quick decisions on whether to invest $500 million in a new oil refinery. Or to the nation, when we examine detailed strategies.

For our purposes, however, an overall and not necessarily a very precise view will be sufficient to show the seriousness of the problem. We will convert the size of a resource into the length of time that resource would last at today's rate of consumption. Our conclusions based on this simple calculation would not have been altered significantly had we chosen a moderately increasing or decreasing rate of consumption.

We can create a "scale" to judge the adequacy of oil resources by noting the U.S. consumes about 17 million barrels of oil each day, or a little over 6 billion barrels each year. (One barrel equals 42 gallons.) For the lower 48 states, our onshore reserves are 40 billion barrels; with Alaska added, the total is 56 billion barrels.[3] The duration of the supply is given by dividing 56 by 6. Thus, without imports, offshore oil or new discoveries, our oil would last about nine years. If this were the full story, we would be in deep trouble. Are we?

As we delve more deeply into the problem it becomes more complex, especially because estimates of undiscovered recoverable resources differ widely among experts, and have undergone radical changes with the passage of time. *Since the overall conclusions expressed in this book rest heavily on trends and interpretations of fossil fuel resources,* we will discuss these questions in more detail. We begin with oil imports.

In 1974, the U.S. imported about 40 percent of its crude oil; that was 16 percent more than during the preceding year.[5] During one week in March 1976 oil imports exceeded 50 percent for the first time! It was an isolated dramatic excursion from a more gradually worsening norm, but, like the first chilling wind in autumn, may be considered to be a forerunner of things to come.

Were we to abandon the thought of energy independence, and maintain a 50-50 split between domestic and imported oil, and were our consumption to remain constant, the above time span would be increased from 9 to 18 years. We also should add to the onshore reserves which we have been considering offshore reserves of 6 billion barrels and, using the best available estimates,[6] another 82 billion barrels of "undiscovered recoverable" oil (both onshore and offshore). This gives a grand total of 144 billion barrels—enough for about 24 years, at our present use rates, without imports, and enough for twice that time if we rely on imports for half our oil.

Let us examine further the meaning of the term "undiscovered recoverable." How do we know it will be discovered? What will be the cost, even if it is discovered? An undiscovered recoverable resource is one geologists believe will be found. Although based on the best geological evidence available at the time the estimate is made, such finds are anything but a sure thing. A bit of insight into the uncertainties may be gleaned from the following facts:[7]

Oil companies have recently drilled about 65 wells off the coast of Newfoundland without finding significant amounts of oil. The cost was $200 million. It was expended in a region that is supposed to contain undiscovered resources. They are still undiscovered. Of course, the very next attempt at drilling may be successful and a vast new oilfield might be discovered. But our reliance on that possibility in planning a course of action involves considerable uncertainties.

To compound the problem, the numbers for undiscovered resources are disputed by many expert geologists, and the various estimates differ widely. In fact, it is just this disagreement which is one of the roots of the fight over nuclear power, because it encompasses the question of need.

If a calculation based on the *most* optimistic estimates[8] for undiscovered recoverable oil is made as before (but with imports left out because they can't be counted on over long periods), we find about 60 years as the length of time we can continue using oil at the present rate. Sixty years may be a long enough period to permit some persons to be complacent about the problem— particularly persons who do not think about the needs of their great-grandchildren. But most others should be alarmed by that short a time scale.

Although 60 years is a very short time scale if oil were our only source of energy for the places where it is now used, it is a sufficient length of time to realistically count on hopes that new methods of energy production, which are still unproven technologically and(or) economically, will indeed be developed in time to meet the demise of oil.

Thus, a crucial question arises: Which are right—the pessimistic or the most optimistic time estimates?

In the preface to this book, we remarked that early in 1973 we became "disturbed" over oil resource limitations. One thing that disturbed us was that in our judgment the high estimate of 60 years was too optimistic. A fact of great significance, which lent credibility to the more pessimistic predictions, as exemplified by Hubbert's analyses,[9] was that the daily U.S. oil production dropped by 4 percent between early 1973 and early 1974 to about 8.9 million[10] barrels a day, suggesting quite directly that America was beginning to run out of domestic oil.

Others have attributed the decreasing rate of oil production to oil company policies, and have rejected the more pessimistic evaluations of resources. For example, in a series of articles in *The New Yorker* magazine, Barry Commoner[11] states that "One of the curious aspects of these disagreements is how sharply they divide oil companies and government agencies." He cites a Department of Interior survey which reports five oil company estimates of undiscovered recoverable oil, all of which give low numbers, and four U.S. Geological Survey estimates, three of which (excluding Hubbert) give high numbers. Placing more confidence in the U.S. Geological Survey, or at least in their high estimates, than in the oil companies, Commoner concludes that "some 325 billion barrels of domestic crude oil are available to us."[12]

However, it is hard to maintain this optimism in the light of the most recent U.S. Geological Survey Study,[13] which in effect repudiates the earlier surveys cited by Commoner. This latest U.S. Geological Survey analysis concludes that the most probable value for the resources is, as mentioned above, 144 billion barrels. Allowing for uncertainties, it places the true number as ranging between 112 and 189 billion barrels.

Another recent independent study was carried out under the auspices of the National Academy of Sciences.[14] It arrives at an estimate of the undiscovered recoverable oil which is only 15 billion barrels higher than the U.S. Geological Survey average.[15] If we were to adopt this slightly more optimistic value, we would have resources of about 159 billion barrels. This is still less than half the value suggested by Commoner.

In summary, some of the early optimism regarding fossil fuel resources was based on estimates made by the U.S. Geological Survey[16] in 1972. Our "60 years" is derived from these estimates. But in the dramatic turn we have mentioned, the U.S. Geological Survey[3] revised oil and gas resource estimates by large downward shifts—by over a factor of four in the case of undiscovered oil. The revision brought USGS estimates into agreement with oil company values. The Survey assigns only one chance in 20 that oil in amounts exceeding the limit of 189 barrels (a 30 year supply) will be found!

Obviously, energy strategy cannot be based on a *hope* as low as one in 20! And oil imports, at a 50 percent level, have an unknown probability of continuance, and should not be counted on to expand the 30 years. Even if they did continue, the cost would be astronomical: 30 billion dollars a year or almost 2 trillion dollars total (if the price is as low as $10 per barrel).

We therefore are led to the overall conclusion that the U.S. certainly should not base any long-range plans on the use of oil for fuel. The seriousness of the situation is compounded by America's dependence on oil for many uses other than as a fuel. It is needed for lubrication, petrochemicals, and plastics, too.

The worldwide oil picture is not much brighter. Middle East reserves will be exhausted in 50 years or less, depending upon the rate of development of the lesser-developed countries. To indicate the effects of increased population and assumed rapid development of so many countries which would like to emulate the U.S., we can make the following calculation: Using M. King Hubbert's[17] estimate of world oil resources, and 5 billion people all using oil at the present U.S. rate, we obtain a time span of 10 years!

As pointed out earlier, natural gas and oil occur in similar geological formations. Reserves of natural gas measured in energy units are about one-half those of oil. Since we use natural gas at about one-half the rate for oil, the picture is not better. One factor makes natural gas reserves worse: gas is harder to import.

The third major fossil fuel is coal; it supplies about one-fifth the total energy consumed in the U.S. The largest fraction is used to generate electricity. That was not always the case. Many of us recall having to stoke the coal-burning furnaces in the basements of our homes in bitter-cold weather when we were young. It was a messy task, and the day an oil-burner was installed as a substitute was an extremely pleasant one.

The energy content of geological resources of coal in the U.S. is considerably greater than that of petroleum and natural gas.[16] Moreover, despite significant uncertainties, the estimates of U.S. coal reserves are probably more accurate because they rest on better-founded geological information than the estimates for oil and gas. As a consequence, not much disagreement has arisen over the magnitude of coal resources; certainly we don't have the dispute that attends estimates on oil resources.

However, there are substantial variations in estimates of how long coal can supply our energy needs, because it is hard to judge how good we will be in 100 years in extracting the relatively less accessible coal deposits. If we assumed that *all* our present energy consumption were to be shifted to coal—and we included only coal deposits already known to exist and readily recoverable through present-day deep mining or strip-mining techniques[18,19]—the time span would be about 130 years. The reason we assumed all energy to be supplied by coal is that, despite some disagreements, oil and natural gas cannot make significant contributions to this longer time span.

At first thought, 130 years seems like a long time, but in the totality of mankind's history on this planet, it is a very short time indeed. Thus, as so aptly expressed by M. King Hubbert[17]—"The entire epoch of the fossil fuels can be only a transitory and ephemeral event in human history—an event, nonetheless, which is unique in geological history and which has exercised the most drastic influence experienced by the human species during its entire biological history."

The debate over fossil fuel resources inevitably leads to economic issues. For example, if the price of oil is high enough, then secondary or tertiary recovery techniques designed to "squeeze" more oil out of otherwise exhausted oil fields might become economically worthwhile. More than half of the oil remains in the ground under present methods of recovery, so the possibilities would seem to be worth exploring.

One reason coal regained a favorable economic position with respect to oil in recent years was the development of strip-mining, which now accounts for about half of U.S. production. However, less than 10 percent of coal is close enough to the surface for strip-mining techniques.[19]

Submarginal and undiscovered deposits[4,18] of coal might raise our time span to as high as 500 years. But to recover these might require desperation digging.

Whatever estimates of coal accessibility are used, it is seen that coal could give the U.S. a reprieve from energy problems for a very long time, a century or more. This would seem to afford ample time to develop radically new energy sources, including, possibly, breeder reactors, fusion, or large-scale solar power. The objection to this course is that heavy use of coal creates environmental problems whose solutions are neither technically nor economically obvious at this time. Some aspects of these problems are worldwide, in that the pollution we create could have far-reaching effects on people throughout the world, as well as in the U.S. This part of the problem will be discussed again in Chapters 17 and 18.

Two other sources of oil must be considered, one from tar sands and the other from shale. The former is found mainly in the north-central regions of Canada and the latter (called oil-shale) chiefly in the states of Colorado, Utah, and Wyoming. Neither plays a significant role in supplying energy today, yet the deposits of these energy sources are extremely large. If environmentally acceptable technology or any commercially feasible technology can be found to utilize their potential, they could extend our use of fossil fuels by many years.

The tar sands of Canada are being commercially studied now, and expectations for successful utilization are high.

Oil-shale containing more than 25 gallons (0.6 barrel) of oil per ton of rock is estimated to be equivalent to about 750 billion barrels,[18] or enough for about 120 years at the present U.S. rate of oil consumption. We note, however, that 25 gallons per ton of rock means only about 10 percent of the rock by weight.

Substantial efforts are now being expended to extract oil from oil-shale in an economic, ecologically viable manner. If these efforts are successful, the useful life of gasoline-powered automobiles will be substantially extended. However, a recent technological review[20] of the methods being considered leads to some discouragement over the prospects. Nevertheless, the effort is worth making because the stakes are so high.

Our simple calculations, based on the time U.S. fossil reserves can last at present rates of consumption, can be summarized for the U.S. as follows:

- The future—even the relatively near future—appears bleak indeed for petroleum and natural-gas resources.
- Coal can fulfill our needs at the present rate for a long time, so that were nothing else available, the U.S. could rely heavily on coal in the next century or so and hope for some radically new sources in the distant future.
- Technological prospects for new sources from oil-shale are not good enough to be counted upon.

Most of the foregoing discussion concentrated on the fossil-fuel picture in the U.S. What is the world outlook? Although it is reasonable to assume a relatively constant future rate of energy consumption in the U.S. and much of the Western world, primarily because that part of the world enjoys a high standard of living, we cannot expect such constancy with the lesser developed and the developing nations. These nations comprise about two-thirds of the world's population. The hopes and desires of these two and two-thirds billion people for better living standards rest heavily on greatly expanded use of energy.

Thus, the demand for fossil fuel may exhaust reserves even more rapidly on a worldwide basis. Indeed, it is significant that the annual rate of increase in energy consumption for the world as a whole over the past 10 years has been about 5 percent, while the U.S. rate has been about 4 percent. This difference can be expected to increase.

The world reserves of oil are concentrated in the Middle East. U.S. reserves are only about one-fourteenth of the world total. Yet the U.S. accounts for nearly one-third the world consumption.

In sharp contrast, almost one-third of the world's coal reserves are located in the U.S.[21] Some persons may view this geological happenstance as fortunate, but it imposes moral responsibilities on the U.S. to neither squander the coal nor use it in an environmentally damaging way.

REFERENCES

1. M. King Hubbert, *Energy Resources,* in *Resources and Man,* W. H. Freeman, p. 166 (1969).

2. See, for example, Ref. 3, p. 8, or Ref. 4, p. 334.

3. B. M. Miller et al., *Geological Estimates of Undiscovered Recoverable Oil and Gas Resources in the United States,* U.S. Geological Survey Circular 725 (1975). Prepared for the Federal Energy Administration.

4. *Mineral Resources and the Environment,* National Academy of Sciences, Washington, D.C. (February, 1975).

5. P. H. Abelson, Science *185,* 309 (July 26, 1974).

6. Ref. 3, pp. 28-29.

7. R. Gillette, Science *185,* 130 (July 12, 1974).

8. See, for example, Table I, Science *185,* 128 (July 12, 1974), citing U.S. Geological Survey, high value.

9. See Ref. 1, p. 186.

10. Robert Gillette, Science *185,* 127 (July 12, 1974).

11. Barry Commoner, The New Yorker, Feb. 2, 1976, p. 53.

12. Ibid., p. 56.

13. See Ref. 3. These results have been widely reported in the press, for example, by United Press International, May 10, 1975 and June 20, 1975.

14. See Ref. 4, and summary in Science *187,* 723 (Feb. 28, 1975).

15. The NAS estimates undiscovered recoverable resources of oil and natural gas liquids of 113 billion barrels (Ref. 4, p. 90). The USGS estimates the natural gas liquids to account for 16 billion barrels (Ref. 3, p. 45). The difference between the two estimates is therefore $113-16-82=15$.

16. P. K. Theobald, S. P. Schweinfurth, and D. C. Duncan, *Energy Resources of the United States,* U.S. Geological Survey Circular 650 (1972).

17. M. King Hubbert, *Man's Conquest of Energy,* in *The Environmental and Ecological Forum,* USAEC Office of Information Services, TID 25857.

18. *Future Energy Outlook,* Quarterly of the Colorado School of Mines *68,* 83 (April, 1973). Also see Ref. 16.

19. Edmund A. Nephew, *The Challenge and Promise of Coal*, Technological Review, Vol. 76, p. 21, Dec. 1973. Reprinted in *Perspectives on Energy*, Oxford University Press (1975).

20. E. W. Cook, *Oil-Shale Technology in the U.S.*, Fuel *53*, 146 (July, 1974). Plans to construct an oil-shale plant in Colorado apparently have been suspended. See Newsletter, Environmental Defense Fund, Nov. 1974.

21. *Energy Perspectives*, U.S. Dept. of the Interior, U.S. Government Printing Office, Report Number 024-000-00812-6 (Feb. 1975).

RENEWABLE RESOURCES
Chapter Four

Solar energy is the most conspicuous renewable energy source. The quantity of energy arriving on earth from the sun daily is tremendous compared to the amount we need. That explains quickly and handily why thermal pollution from power plants does not change the earth's temperature enough to be a problem, except in very specific localities. It also suggests that the harnessing of solar radiation could solve all our energy needs in perpetuity—*in principle*.

Simple numbers bear out the statement. The average amount of solar power falling upon one square yard of an Arizona desert is about 120 watts.[1] It has been projected that the total electric-power capacity in the U.S. in 1980 will be equal to or somewhat less than 600 billion watts.[2] Assuming for the moment 100 percent efficiency in conversion of solar power to electric power, it follows the solar power falling upon 5 billion square yards, or less than 2,000 square miles, would provide all the U.S. needs for electric energy in 1980. That may sound like a considerable area, but it corresponds to a square less than 50 miles to a side.

If we were to allow for the unavoidable inefficiency of the energy conversion process, we more realistically would need an area several hundred miles to a side, but still far from prohibitive from the land-use standpoint.

Obviously, solar energy falls in varying amounts over the entire country. Power from the sun is already used in innumerable ways. All fossil fuels owe their original source of energy to the sun, although they are not a renewable source on a reasonable time scale.

On a renewable basis, an exploitation of fast-growing vegetation offers some prospect of converting solar energy to usable power, but the efficiency thus far is relatively low. Indeed, over one billion people in the world now depend almost entirely on firewood for their energy needs. They use one-half of all the wood cut in the world, and it is becoming more and more scarce. These people face a grim future.[3]

Other indirect uses of solar energy are found in the production of hydroelectricity (the water coming from rain, which, in turn, depends on evaporation of surface water), of wind power, and of power from temperature differences in the ocean.

Hydroelectric power has become quite important in the U.S., generating about 16 percent of the nation's electricity.[4] But there is little possibility of significant expansion. Total world hydropower resources can be estimated with high accuracy. If every stream and river were utilized,[5] enough electricity could be produced for about 1.5 billion people at the rate of consumption now enjoyed in the U.S. In some geographical regions, the lakes created by hydro-dams are a definite asset to the landscape, but in others they may be irreversible ecological disasters.

(An example of a disaster is the dam in Glen Canyon in Arizona. John McPhee recounts the details in *Encounters With the Archdruid*, Farrar, Straus, and Giroux, New York, 1971.)

People have designed modern windmills which are more efficient, though less picturesque, than old windmills, but estimated costs are high (except for isolated places), and expectation of a major contribution from windmills does not appear realistic.[6]

In certain ocean regions it might be feasible to utilize solar energy because of the differences it generates in the temperature of the water at the surface and at a great depth. The scheme involves a heat engine to drive electric generators,[7] but because of the relatively small difference in absolute temperatures, the engine would have a very low thermal efficiency. It is quite possible the method will work, but it remains to be shown that such power plants are commercially viable. Proposals envisage energy conversion in the water to hydrogen gas, which could then be brought to shore. At present, the methods for direct use of hydrogen in the energy economy are not at hand, but they are probably technically feasible.

The most important potentials for solar energy lie in its direct utilization. They fall into two general categories: Small-scale use at the site of consumption, such as rooftop building heaters, and large-scale electric-power production in desert regions of the U.S.

An ambitious research and development effort—more than $160 million was requested by the administration for ERDA alone in 1976-77 for solar energy[8]—is underway on systems for heating or cooling of buildings and for heating water. In Southern and Eastern regions of the U.S., the greatest energy savings would be in home heating systems and cooling systems in the summer months. In fact, the present electric energy consumption in homes is often higher in the summer than in the winter because of the great energy requirements of air-conditioning units. Solar cooling systems employ liquids which produce cooling when rapidly evaporated.

Solar heating panels for roof-top installation are already commercially available. Unfortunately, their cost is rather high. Since their main components are glass and sheet metal—techniques for whose manufacture have

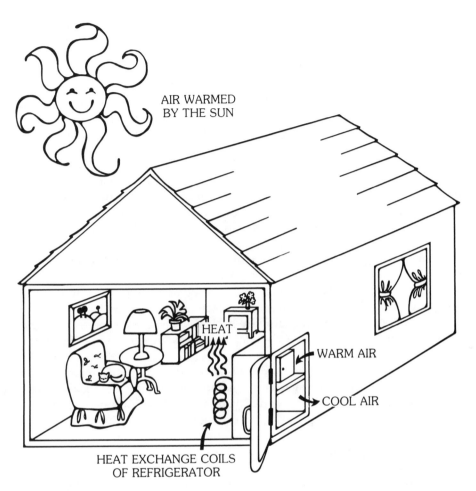

AIR WARMED
BY THE SUN

HEAT

WARM AIR

COOL AIR

HEAT EXCHANGE COILS
OF REFRIGERATOR

Illustration of a heat pump. A heat pump is really a refrigerator used in reverse. It "refrigerates" the outside air and heats the house. You can feel the heat coming from your home refrigerator when it is running. Heat pumps for home heating are just refrigerators designed for the purpose.

been around for a very long time—it is unlikely that any drastic reduction in cost can be achieved.

Solar panel heating systems must include a certain amount of plumbing equipment, including circulation pumps. These are most conveniently operated by electricity. The plumbing equipment must include sufficient storage capacity to provide heat in the evening and night time.

Of course, a back-up heating system is generally necessary to tide over a period of inclement weather. Again, electric heating is convenient for this purpose.

Calculations and experiments indicate 50 to 70 percent of the heating requirements of a typical American home can be supplied by solar heating panel systems.

A drawback is that electric power requirements for a city heavily equipped with individual solar heating units must be sufficient to handle very large "peaking" loads whenever the weather is poor. Capital investment costs for both individuals and power companies would thus be high.

An alternative source of solar energy is air—that is, the atmosphere, which is heated by the sun. It retains its solar heat for long periods, is available during both day and night during all weather conditions, and is free for everyone. The trouble is that the heat is too "low grade"; that is, it is at a low temperature. However, the heat of the atmosphere can be extracted and raised to a comfortable temperature by means of heat pumps. The energy input required to "elevate" the atmosphere's heat content to a reasonable temperature ranges from about 30 to about 65 percent of the heat derived from the atmosphere. Thus, heat pump systems are in the same range of efficiency as solar-heating panel systems.

Solar heating by means of heat pump systems does not suffer from the "peaking load" drawback of solar-panel systems. Moreover, heat pumps are available on a very wide scale and are backed by long-standing technological development. The yellow pages of telephone directories of most cities abound with vendors eager to install them.

Like the auxiliary equipment and backup systems for solar-panel heating methods, heat pumps are most conveniently powered by electricity. In common with almost all schemes for reducing energy consumption, heat-pump systems are expensive to install. Nevertheless, they are already competing favorably with more familiar forms of home heating in many geographical regions, because of the savings in fuel costs.

However, as pointed out by Hans Bethe,[9] residential heating accounts for only about 10 percent of our total energy use. Thus, if even by the year 2000 one-third of our homes were to cut their heat-energy demand in two by using solar panels or heat pumps, the savings would be less than 2 percent of our total national energy budget. This is a worthwhile saving, but hardly decisive.

Turning to the other major approach for using solar energy, proposals involving methods for desert power plants remain in the idea stage for the time being. Arizona is often mentioned as the most favorable site for such energy factories.

A typical large power plant produces about 1,000 megawatts, which is sufficient to provide the electricity needs of a city of more than half a million people. To produce that much power from a solar-power station at 10 percent efficiency would require about 100 million square yards, or a 30-square-mile area. That is roughly the size of a large city itself. Once again, it can be said that is not an excessively large area, given the massive acreage of the U.S. However, that is not the principal problem.

The technical problems associated with a vast array of suitable solar collectors, which still must be developed, may be solved some day. But as yet no economically feasible method for doing it has been found.[9]

In addition to the huge problem of developing satisfactory collectors, other serious technical problems are foreseen:

- Sunshine is not constant, thereby necessitating a method for storing the energy.
- Present-day techniques for transmission of electric power are not particularly feasible beyond 1,000 miles or so.
- The solar collectors might be quite vulnerable to storm damage, to dirt, or to sabotage; even with optimal collectors, the economic costs may be excessive.[9]

Despite all these apparent disadvantages, research in direct conversion of solar energy to electric power on a large scale is worth a try. Again, however, it is unwise to gamble the future of the nation and the world on the uncertain probability of success.

Geothermal energy is another—at least semi-renewable—energy source receiving a great deal of popular and research interest. One method utilizes existing steam sources, such as exist in Yellowstone National Park. A small contribution to the total needs of the U.S. probably can be derived from such sources and a few plants already exist,[10] but possibilities for large-scale expansion seem remote.

A second geothermal scheme receiving considerable research money and support is to utilize the heat of the rock of the earth at great depths. It is referred to appropriately as the "hot rocks" method.[11] Although it is worth pursuing on a research basis, success there also lies in the "maybe" realm, and serious reliance cannot be placed on solving the U.S. energy problem in this way.

REFERENCES

1. One finds various numbers quoted, depending on the source. The exact value is immaterial. Hubbert (Ref. 17, Chapter 3, p. 30) gives 145 watts.

2. There are many such "projections." The 600,000 megawatts is a projection made by the AEC in WASH-1250, Appendix I, 1-4 (July, 1973).

3. New York Times, Oct. 6, 1975. (Summary of a report from Worldwatch Institute, Washington, D.C.)

4. *Energy Research and Development—Problems and Prospects,* p. 29. U.S. Govt. Printing Office (1973).

5. Based on Hubbert, p. 29, Ref. 17, Chapter 3.

6. Northwest Public Power Bulletin, Feb. 1974.

7. C. Zener, Physics Today, p. 48, Jan. 1973, and *Physics and the Energy Problem,* p. 412, A.I.P. Conf. Proc. Am. Inst. Phys. 1974.

8. Nucleonics Week, January 23, 1976. (It seems probable that Congress will allocate even more than this amount for solar energy research.)

9. H. A. Bethe, *The Necessity of Fission Power,* Scientific American, Vol. 234, p. 21, (January, 1976).

10. Hubbert, p. 33, Ref. 17, Chapter 3.

11. M. C. Smith, *Physics and the Energy Problem,* p. 401, A.I.P. Conf. Proc. Amer. Inst. of Phys. 1974.

THE POTENTIAL FOR NUCLEAR ENERGY
Chapter Five

As already indicated, the principal thrust of this study is an analysis of nuclear-fission energy, its prospects, its advantages, and the reasons which have led us to conclude it is the best—and possibly the only—viable alternative.

The initial focus is on the resources of various kinds of nuclear energy, paralleling the format used for fossil-fuel resources. It is important in understanding nuclear energy to begin with its basic tenets.

In principle, the light-weight elements, such as "heavy" hydrogen (or deuterium), lithium, and boron, can be converted into slightly heavier elements with the release of enormous amounts of energy. Although the natural fraction of heavy hydrogen in ordinary hydrogen is only 16/1000 percent, the ocean waters contain vast quantities of heavy hydrogen, so there is no question concerning adequate long-term supplies of this energy resource.

Synthesis of light elements into heavier elements is referred to as nuclear fusion. The technical problem there is to find a way to perform the fusion in a controlled (slow) manner. At the present time, all we know is how to do it in a rapid, or explosive, manner—as with the hydrogen bomb. Thus, it is known how one might be able to use fusion energy, but the technical achievement of controlled fusion has not yet been accomplished.

The heaviest natural elements, particularly uranium and thorium, also can be made to release enormous quantities of energy. (See the conversion chart, Appendix A.) This occurs when their atoms are "fissioned" or split into two roughly equal halves, resulting in atoms of medium-weight elements. Although other heavy elements, such as lead, would produce energy in "fissioning," it is only possible to produce the process in a useful manner with elements beginning with uranium and thorium.

Uranium and thorium are plentiful elements; they are found, for example, in ordinary granite. Uranium constitutes about 3 parts per million of average earth and rock; and about 5 parts per million in granite;[1] even coal contains uranium in the ratio of a few parts per million.[1,2] This may appear to be an insignificant quantity, but it turns out that the energy content of the uranium in coal is therefore greater than the energy gained in burning the coal! (It is assumed here that all the uranium is useful as an energy source; later it will be explained under what conditions that is the case.)

It has been estimated that the low-grade deposit of uranium in Tennessee, known as Chattanooga Shale, which has a concentration of about 60 parts per million, contains more than 5 million tons of uranium.[3] Only 2 percent of the area of Tennessee thus contains as much energy as all the fossil deposits known or suspected in the U.S., and this includes the sum total of oil, gas, and oil-shale! Two percent of Tennessee is about 625 square miles.

By way of comparison, an area ten times greater or about 6,200 square miles—that is, a square of about 80 miles to a side—has already been strip-mined for coal. About half has been reclaimed, but about 7 more square miles are strip-mined each week.[4]

The Chattanooga Shale is not the source of uranium presently utilized by the U.S. A much richer but smaller deposit lies on the Colorado Plateau.[5] For reasons to be detailed later, only the use of relatively rich ores is economically feasible for energy production at the present time.

Another vast store of uranium resides in ocean water,[6] the energy content of which would serve man's needs for about 180,000 years,[7] even if everyone on earth today used energy at the present U.S. rate per capita.

Thorium deposits rival those of uranium in total energy content. To summarize, we are thus led to the conclusion that as a source of energy, uranium and thorium can satisfy man's needs for a period of time beyond easy conception.

If the means for converting the energy of uranium and thorium to a useful form were as simple as those for fossil fuel, the world's energy problem would be solved. Unfortunately, that is not the case. A number of "catches" prevail.

Understanding the first "catch" requires another quick discussion of physics. Uranium atoms in the raw consist of two different kinds of nuclei, or isotopes, each with slightly different masses and structure. They are referred to as U-235 and U-238, respectively, where the numbers are the same as their relative nuclear weights.

Natural uranium is only 0.7 percent U-235; more than 99 percent is U-238. Only the U-235 will readily "burn" in a chain reaction in which the "chain" is sustained by neutrons passing from nucleus to nucleus. Coal burns by a chain reaction in which the "chain" is sustained by heat energy passing from atom to atom. Thus, the neutrons play the role of heat by causing U-235 to undergo fission and producing more neutrons in the process.

As a result, the U-235 is referred to as "fissile" because it can be fissioned. The U-238, on the other hand, acts chiefly like a strong damper by absorbing neutrons and preventing the chain reaction from proceeding, as would occur if sand were mixed with coal. This is the reason naturally-occurring uranium does not "burn" spontaneously.

Unlike the coal-sand mixture, however, the U-238 exhibits a most remarkable property when it swallows up neutrons: It produces, by way of a time-delayed radioactive chain of decay, an element which does not exist in nature but which is itself capable of sustaining a chain reaction, just as for U-235. It is called Plutonium 239, or Pu-239. For that reason U-238 is labeled a "fertile" isotope because it can produce a fissile substance.

Although natural uranium cannot in general produce a chain reaction, a specially designed lattice-like structure, made of uranium metal and interspersed with a light-weight element that does not itself absorb neutrons, will allow the U-235 component to react spontaneously and produce heat. That is a rather rough description of an "atomic pile." Such piles were constructed for the Manhattan Project during World War II. The first man-made release of nuclear energy, achieved in a pile in December, 1942, was thus slow and controlled. (The atomic bombs, built several years later, needed much purer fissile material.) Today, particularly in Canada, some commercial reactors of this type using natural uranium are used to produce electric power.

The neutron-absorbing capability of U-238 has an important consequence, not only because the Pu-239 generated is itself a fissile material like the U-235, but because the plutonium, being a different chemical element, can be separated from the uranium by relatively simple methods. The U-238 becomes useful by converting it into Pu-239, and therefore the abundant U-238 can be classified as an energy source. By way of example, the time-span estimate for the uranium content in ocean water would have to be reduced from 180,000 years of energy supply to 1,200 years if conversion of U-238 to plutonium did not take place. That's quite a hefty difference.

Although the nuclear chain reaction can be achieved in a properly arranged large structure consisting of natural uranium and a suitable "moderator," such as carbon or "heavy" water to slow down neutrons, an alternative arrangement using ordinary or "light" water as a moderator can be made to react, provided the uranium is somewhat "enriched" in the U-235 isotope. The reason is that ordinary water absorbs some of the neutrons and acts as a damper.

The trouble with using light water, which is much much cheaper than heavy water, is that an isotope separation system is required. The first such plants for enriching uranium—the "diffusion" plant and the "electromagnetic" plant—were built in the Second World War. Their primary purpose was to enrich uranium to nearly pure U-235, so that bombs could be made.

(Of the two types, most of the wartime U-235 was produced by electromagnetic separation. The diffusion plant was just beginning operations at the end of the war. Later, more diffusion plants were constructed.)

Illustration of a chain reaction in uranium. These reactions produce energy, Pu-239 and Pu-240, as well as fission products. The pictures are arranged in time sequence as imaginary snapshots.

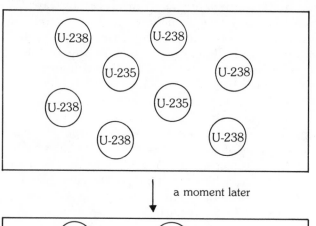

A mix of uranium atoms. In LWR fuel there are about 30 U-238 atoms to one U-235.

a moment later

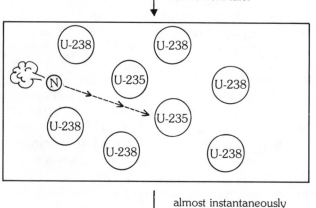

A moving neutron enters and is absorbed by a nucleus of U-235

almost instantaneously

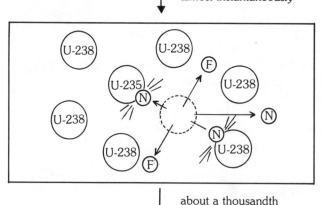

The atom of U-235 disintegrates releasing 3 neutrons (N) and 2 fission fragments (F). One fragment is usually larger than the other.

The neutrons collide first with nuclei of the moderator (not shown) and then with nuclei of U-235 and U-238.

about a thousandth of a second later.

(Continued on next page)

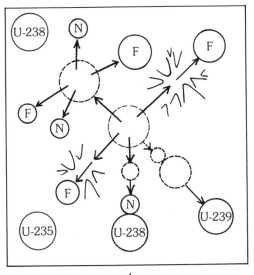

When one of the released neutrons collides with more U-235, it in turn disintegrates and releases more neutrons and fission fragments.

When a neutron collides with U-238 it is "swallowed" and the atom of U-238 becomes U-239

The fission fragments slow down, releasing energy as heat.

The process continues on and on in "chain" fashion, always sustained by more neutrons.

About | 20 minutes later

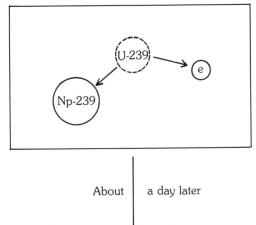

U-239 "decays". An electron is released, and a positive charge is "born" in U-239, thus converting it into Np-239. Another particle, called a neutrino (not shown), is also released.

About | a day later

(Continued on next page)

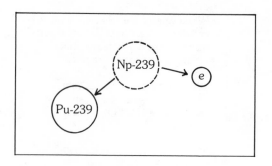

Np-239 decays to Pu-239

As time goes on

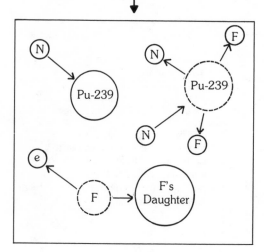

Some Pu-239 absorb neutrons and become Pu-240, and some undergo fission. Old fission fragments decay and make new isotopes (daughters).

Nuclear-power plants of the light-water type most common in the U.S. use uranium enriched from the 0.7 percent natural value to 3 percent or 4 percent in the immediate fissionable U-235 isotope. The enrichment procedure reduces the capital cost of the power plant but increases the cost of the fuel, as compared with natural uranium.[8] At present, the existing U.S. diffusion plants are operating at virtually full capacity. They will supply enriched uranium to reactors for about 72 U.S. electric utilities and to reactors in about 18 foreign countries.[9]

It might be inferred that it is a simple matter to burn U-235 in a nuclear reactor and create Pu-239. Actually, the present-day light-water reactors (LWR's) do, indeed, produce about one-half as much Pu-239 as the U-235 they consume, and about one-half of that Pu-239 is then burned in the same reactor, so that "spent" fuel finally contains about one-quarter as much Pu-239 as the original U-235. But to convert U-238 into a long-term energy source, the amount of Pu-239 produced must at least equal the amount of U-235 burned so that the stock of fissile material is maintained. The equality must be retained over the entire fuel cycle, including any losses incurred in the chemical separation of plutonium from the original fuel.

As time goes on, the world would thus eventually exhaust the supply of U-235, but Pu-239 would be there to continue the "breeding" cycle. The recycle time for the fuel must not be too long, for if it remains too long in storage, the effective breeding rate, which is analogous to the interest rate on a savings account, may become too small to "keep up" with energy demands.

The process sounds easy, but it is not. Reactors capable of performing this breeding process, appropriately called "breeder reactors," are not yet available as commercially viable items, at least in the U.S. Experimental breeders have been built in the U.S.—actually the very first reactor to produce electric power was a breeder—and a demonstration commercial breeder reactor is in the design stage. Foreign breeders exist. One in France, called Phenix, is already supplying power to the French "grid."[10] The U.S. demonstration breeder may be built in Tennessee; nevertheless, it appears at this time that European countries are ahead of the U.S. in breeder development. A more extensive discussion of breeders is reserved for Chapter 14.

Commercial reactors now in operation in the U.S. (about 55), most of which are Light Water Reactors, and those planned or under construction (another 100 or so) are not capable of producing enough Pu-239 to use very much of the vast U-238 resources. They use about 0.6 percent of the original uranium by burning U-235 and another 0.3 percent is used by internal production and consumption of Pu-239. If the Pu-239 recovered from spent fuel is recycled, the LWR's can use about another 0.3 percent of the total uranium resources.

Therefore, present-day reactors are able to use only about one percent of the energy available in natural uranium. It is important, however, to note that the spent U-238 fuel is not lost. It can be stored for later use in breeders.

The second "catch" to easy, full realization of the vast potential of the uranium concerns economic problems in the utilization of low-grade uranium sources for the non-breeder-type reactors. At the present time, the cost of uranium itself is a small fraction of the total cost of electricity produced from uranium. As of 1973, with the price of uranium at $8 to $10 a pound, fuel costs were less than 1 mill per kilowatt-hour, provided the plutonium generated is ultimately recycled.[11]

Thus, the U.S. can afford to use considerably more expensive uranium, and still fission power will be economically advantageous, as compared with coal and oil—particularly if the prices of the latter continue to rise. However, if the U.S. is forced to use uranium ore of low concentration—

such as the Chattanooga shale, for example—one of the main environ-mental advantages of nuclear power would be lost because it will then be necessary to process quantities of rock comparable to the amounts re-quired for coal-mining.

It is sometimes argued that the supply of natural uranium in the United States is too small to justify the present light water reactor program. This is not the case. Recent estimates of uranium resources indicate that we have about 3.4 million tons of uranium oxide, for ore rated at "$30 per pound."[12] It takes a little over 200 tons per year (without plutonium recy-cle) to fuel a 1,000 Mw(e) reactor. Thus 3.4 million tons are sufficient for about 15,000 years of reactor operation.

This fuel would be economical to use. Even if its actual price doubled, to $60 per pound, the costs of the fuel would amount to only about 5 mills per kilowatt-hour, or about 10 to 20 percent of the expected electricity price. Oil and coal promise bigger rises. (More information on the relative economics of nuclear power vs coal is presented in Chapter 8.)

Before a reactor is built, there should be a fuel supply available to last for 30 years—its rated lifetime. Thus the 3.4 million tons would suffice for about 500 reactors, each running for 30 years. The most optimistic recent plans, which now seem unlikely to be fulfilled, called for 250 reactors by 1985. Conceivably we could reach 500 reactors by 1990. This eventuality seems unlikely, but might be accomplished with a change in public mood and vigorous government policies.

Should that happen, we could "run out" of uranium in 1990. But we would only "run out" in the sense that we would not have the resources, without further discoveries or imports, to warrant the building of more uranium fueled light-water reactors. The existing ones would keep going, most of them past the year 2115. During this period they would represent a major energy source, providing considerably more electricity than we now obtain from all means of electricity generation taken together.

However, if ordinary light-water reactors are to provide power for much of the 21st century, it will be necessary either to find new reserves of reasonably high-grade uranium (quite possible, but not to be counted on) or to demonstrate the practicality of methods of utilizing uranium from low-grade sources, such as exist in Chattanooga shale or, more remotely, in seawater.

In view of the uncertainties in the prospects for more uranium, it is most prudent to view the non-breeder fission reactors as a valuable intermediate-term energy source, but for long time scales to plan on the breeders. Thoughtful persons will rest more easily when it is known with certainty that the technical and economic problems of breeders have been solved.

At least one economic problem with breeders, however, has already been solved—the cost of fuel per energy unit is so low that it is feasible to use very low-grade uranium ore.

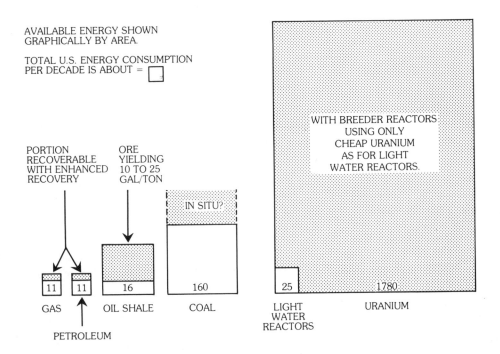

AVAILABLE ENERGY SHOWN
GRAPHICALLY BY AREA.

TOTAL U.S. ENERGY CONSUMPTION
PER DECADE IS ABOUT =

WITH BREEDER REACTORS
USING ONLY
CHEAP URANIUM
AS FOR LIGHT
WATER REACTORS.

PORTION   ORE
RECOVERABLE   YIELDING
WITH ENHANCED   10 TO 25
RECOVERY   GAL/TON

IN SITU?

| 11 | 11 | 16 | 160 | 25 | 1780 |

GAS   OIL SHALE   COAL   LIGHT   URANIUM
                                       WATER
                                       REACTORS
      PETROLEUM

Summary of fossil and uranium energy resources in the United States. (Based on ERDA-48.) The numbers in the boxes indicate the number of years each resource would last if *all* U.S. energy at the present use rate were supplied by that resource alone. Only the more assured (in the lower boxes) are included.

Thorium cannot be used directly as a fuel in nuclear reactors. It is similar to U-238 in that it must first be converted into a new form, in this case U-233, which also does not exist in nature. The conversion can occur in one type of reactor commercially developed today—the High Temperature Gas-Cooled Reactor (HTGR). However, the conversion efficiency for HTGR's is only about 80 percent; they still use up the rarer U-235 but at a slower rate than the LWR's. Other types of reactors that have been projected may be able to breed more U-233 than U-235 consumed.

In summary, uranium and thorium constitute a substantial resource base without breeders and an enormous base with breeders. Moreover, neither element essentially has any other major use, although we can still extract radium from uranium for the luminous dials on watches and utilize uranium for making such pretty things as colored glass. Obviously, oil and coal have more important alternative uses.

On a long-range, global scale, then, it makes eminently good sense to exploit uranium as an energy source and save oil and coal for the chemical industry and for other critical uses, where they have unique advantages.

REFERENCES

1. R. P. Hammond, *Nuclear Power Risks,* American Scientist *62,* 155 (March-Apr. 1974).

2. M. Eisenbud and N. G. Petrov, Science *144,* 288 (1964).

3. Hubbert, Ref. 1, Chapter 3, p. 227.

4. R. Gillette, Science *181,* 524 (Aug. 10, 1973).

5. Hubbert, Ref. 17, Chapter 3, Fig. 25.

6. Bruce Tucker, Nucleonics Week, p. 2 (May 30, 1974).

7. Based on U.S. rate = 70×10^{15} Btu/year, U.S. Population = 2×10^8, World Population = 4×10^9, and 4×10^9 tonnes U (Ref. 6).

8. C. E. Larson, Science *184,* 849 (May 24, 1974). Present cost of enrichment is about 0.5 mills/kwhr.

9. Nucleonics Week, p. 3, Oct. 17, 1974.

10. See Chapter 14 for additional discussion.

11. Cornell Workshops on the Major Issues of a National Energy Research and Development Program, College of Engineering, Cornell University, p. 136 (Revised Ed. Dec. 1973).

12. R. D. Nininger, Transactions of American Nuclear Society, *21,* 241 (1975); *21,* Supplement No. 2, 75 (1975).

RESIDUES FROM ENERGY PRODUCTION
Chapter Six

The burning of coal produces carbon-dioxide gas and a residue of ashes, some of which escape up a chimney or are caught in suitable filters. Paradoxically, the total weight of the residues exceeds the initial weight of the coal.

Oxygen, which is good for people, combines with the coal to make the carbon dioxide, which can be bad for people but is good for vegetation. Coal also has many impurities, principally sulfur; in burning, sulfur dioxide also goes up the chimney, and no doubt exists it is unhealthy.

Uranium present in coal, as well as other radioactive products, including radium, also takes the chimney route out. All these materials are spewed into the atmosphere and they have direct effects on human welfare and death rates. They also may have serious and possibly irreversible long-term effects on the environment. These are discussed in Chapter 17. Oil-fired power plants emit both carbon dioxide and sulfur dioxide, but they do not emit appreciable radioactive materials. Natural gas is the "cleanest" source of energy when burned. Its only real residue is carbon dioxide.

In normal operation, a nuclear plant emits virtually nothing into the atmosphere. It is not quite "nothing," however, and the very small difference between "not quite" and "nothing" is important enough for an extensive separate discussion in Chapter 10. Here we will discuss the origin of these emissions.

The residue, or ashes, from the nuclear-fission process taking place in the fuel elements of the reactor consists of a great deal of radioactive material. When a fissile nucleus, such as U-235 or Pu-239, is fissioned after absorption of a neutron, two medium-weight nuclei plus two or three neutrons are formed and driven apart by great forces. Exactly which elements and isotopes constitute the two pieces cannot be predicted in advance, but there is a known probability of this pair or that pair of nuclei being created.

An analogy might be made with the breaking of chicken wishbones: It is usual that one fragment is heavier than the other, but the exact split is not known until after the event. On a statistical basis, however, when millions of atoms undergo the split, we know how much of each fission product is produced.

As a result, an array of medium-weight elements is produced in the body of the nuclear fuel. Many are rare chemical species, such as lanthanum, promethium, and cerium, but more familiar ones, like cesium, barium, and strontium, also are created. Most of these fission products are radioactive isotopes and emit radiation. Many decay rapidly and turn into stable nuclei but a few have long half-lives, notably cesium-137 (30 years), and strontium-90 (29 years).

Although part of the residue of the fuel elements remains potentially dangerous for many years, after a short time most of the residue consists of stable isotopes and, consequently, represents no danger. The radioactive portion, on the other hand, presents a disposal problem, examined at length in Chapter 12.

When they decay, the radioactive fission products themselves emit considerable energy, most of which produces heat in the fuel elements. If a nuclear reactor is turned off—an operation typically achieved by the absorption of neutrons in control rods made of an element with great affinity for neutrons—the fuel elements will still continue to emit heat that is equivalent initially to about $1/20$th of the operating power level before shutdown. In a relatively short time, the heat dies away to about $1/50$th of that level, but it is still sufficiently great to cause damage to the reactor unless the presence of cooling fluid is maintained. The possible consequences of a loss of coolant are detailed in Chapter 13.

The radioactive residue has one other important characteristic. A few of the radioactive isotopes produce "daughter" isotopes that are most unusual in that they also emit neutrons. (The most familiar kind of decay produces beta particles, or negative electrons.) Most neutrons are emitted simultaneously with the fission event, but these particular neutrons are delayed after the primary fission by times which are characteristic of the half-lives of the "parent" isotopes. Once created, these delayed neutrons serve to sustain the chain reaction in the same manner as the prompt neutrons—those emitted at the instant of fission.

It is the delayed neutrons, with half-lives up to several seconds or more, which make it possible to regulate or control the nuclear reactor by adjusting the relatively slow-moving mechanical control rods. It is most curious that nature provided us with this built-in control mechanism. Without it, reactors would not work. Only bombs would.

Nuclear reactors produce one other source of radioactivity. Some of the neutrons are absorbed by structural materials or by the moderator, such as water, and by impurities in the moderator. These absorption processes can create new isotopes which are themselves radioactive.

Fission products that are the residue from spent nuclear fuel actually weigh about one-tenth of 1 percent less than the weight of original uranium. The lost mass has been converted into energy in accordance with the famous Einstein Mass-Energy Formula:

Energy = Mass times the square of the velocity of light, or $E = mc^2$.

Albert Einstein proposed his Theory of Relativity, which included the equivalence of mass and energy, in 1905. Actually, a similar loss in mass occurs in the burning of coal by oxygen, but for a relatively inefficient fuel like coal, the change in mass is only about one hundred-millionth of a percent, and it is entirely undetectable by ordinary means.

The possibility that radioactivity in the nuclear-fuel residue and in the other structural materials might reach the environment is responsible for three of the most controversial issues surrounding nuclear power—radiation hazards, waste disposal, and reactor safety. Each issue will be addressed in separate chapters.

It should be noted, however, that the effects of the residues are not exclusively deleterious. Some of the radioisotopes have found uses in medical research and medical therapy, as well as in other practical applications.

CHARACTERISTICS OF NUCLEAR ENERGY
Chapter Seven

In taking stock of nuclear energy resources, reference was made to the enormous energy content of the fuels by weight. To help understand what this means, it is worthwhile to make a few comparisons.

A single kilogram (2.2 pounds) of uranium—a cube about 1½ inches to a side—contains the same amount of energy to be found in about 2½ thousand tons of coal, or about the quantity of coal contained in 30 railroad cars, each transporting 80 tons! It is also equivalent to about 10,000 barrels of crude oil, each one of which is equivalent to about ¼ ton of coal.

Thus, a piece of uranium in the form of a cube about 18 inches to a side is equivalent to the 17 or so million barrels of oil the U.S. uses each day. The energy contents of U-235, Pu-239, or U-233 are all very nearly the same. By conversion to Pu-239 or U-233, energy contents of U-238 and Th-232 are the same as for these fissile isotopes.

About 7 kilograms of pure U-235 are sufficient to supply a city of a million people with electricity for a day—roughly the average output of two large (1,000 megawatt) power plants.

If nearly pure U-235,[1] or Pu-239, were used in the power plants, the fuel required could literally be carried by one person a day on his back. He or she would also be able to carry a sleeping bag, tent, and food for a few days, in addition to the load of fuel. A ground cloth would serve as adequate shielding from the minor radioactivity of the fuel.

For equivalent coal-fired plants, several long freight trains each day would be required to carry the coal! The coal trains would preempt a large length of track, and the piece of land for the track would be about 50 feet wide. An enormous area is wasted, not including the sidings at the ends or the storage for the coal. The sole back-packer, on the other hand, would need only a trail 2 feet wide and no storage space to speak of.

Such a comparison is, of course, not realistic because the nuclear-reactor fuel elements should be returned for reprocessing to recover the Pu-239 and to remove the fission products. The spent fuel must be heavily shielded, so our backpacker would no longer suffice. Also, in present reactors we do not use pure U-235 and about 30 times the weight in fuel is needed.

In truth, real nuclear power plants require an average of about one truckload a week for shipping fuel to and from the plants.[2] It is possible that more traffic will be created by delivery of materials for coffee breaks.

Another important advantage is apparent in the high concentration of nuclear energy by weight: A nuclear plant can operate for a long time—up to a year—without refueling, a factor that could be crucial in emergencies, when transportation and communications might be disrupted.

Comparisons like those cited were a part of popular-science literature shortly after the Second World War, and hope spawned that nuclear energy soon would be available for a host of applications, ranging from private automobiles to planes. For example, an enormous effort was actually made to develop a nuclear-powered plane. About a billion dollars was spent in research over a 15-year period before the project was finally abandoned in 1961 as a technological failure,[3] and the example is a strong reminder that technology has limits.

On the other hand, nuclear-power plants for naval ship propulsion, particularly for submarines, have proved very successful and the U.S. Navy had 112 nuclear-powered ships in operation by mid-1975. Despite some economic and political uncertainties, an important potential may materialize for nuclear-powered merchant ships, as well, although the American ship, Savannah, was apparently uneconomical at the time it was built.

At first, the development of nuclear energy to produce electricity seemed to be a rather simple matter. Atomic piles at the Hanford plant near Richland, Wash., had been built in haste to make Pu-239 in the Second World War, and they wasted many megawatts of power that was pouring uselessly into the Columbia River; it therefore seemed reasonable to utilize the heat to make electricity.

The development of commercial power plants, however, proceeded very slowly in comparison with the fast pace of the military side of the Nuclear Age and was delayed long beyond the proof of the feasibility of the basic chain-reaction process. Most of the development was carried out by the U.S. Atomic Energy Commission in its large laboratories, rather than by private industry, although large manufacturing concerns played important roles, just as they had in the Manhattan Project itself.

Government assistance in the realm of nuclear power is similar to the early role the federal government played in the construction of very large hydro-power dams, for example, Bonneville in the Northwest or Norris in the Tennessee Valley. In those cases, the investment was too large to be shouldered by private utilities either singly or in concert. At the time, criticism of the government projects came from the politically conservative sectors of society. Today the conservatives and liberals seem to have exchanged roles, and nuclear energy is criticized for being federally "subsidized."

Although the technical aspects of nuclear power did not delay construction of early pilot plants—the first actually dates from 1952[4]—it was a long time before reactor technology was developed to the point at which nuclear power could compete economically with fossil-powered electric-generating plants. Today the cost of generating electricity by nuclear fission is lower than the cost for fossil fuel, provided the nuclear-power plants are built on a large scale—about 1,000 megawatts of electric power, or enough to support a city of half a million people by U.S. standards.

Thus, apart from ship propulsion, the use of controlled nuclear energy is at present confined to the production of electricity. The difference between a fossil-fueled and a nuclear-fueled power plant lies only in the manner in which heat is generated to produce steam or another driving substance to power the turbines and electric generators. A large portion of the technology, obviously, was already at hand when the nuclear era dawned.

The efficiency with which heat is converted into electrical energy in a steam power plant depends chiefly on the temperature of the steam as it enters the turbine. In turn, the upper limit of temperature depends upon the strength and other characteristics of the construction materials used.

As it turns out, the efficiency of very large, modern coal-fired plants is about 40 percent at most, whereas it is about 33 percent for present-day, operating nuclear plants.[5] For a coal-fired plant, the total heat energy expended is thus 2½ times the useful energy generated; for a nuclear plant, it is 3 times the useful energy. Thus, for the same electrical output, a nuclear plant produces about 20 percent more total heat than a coal plant.

The difference between the total energy used by the power plant and the electrical energy produced is called waste heat, and it constitutes a source of local thermal pollution. Again for the same useful output, the local thermal pollution of a nuclear plant is about 35 percent greater than for a coal-fired plant of equivalent size. Both types of plants may discharge their excess heat into a large body of water—for example, the ocean—or into the air by means of cooling towers. However, a portion (about 15 percent) of the fossil plant's thermal pollution always goes into the air through smoke stacks, or up the chimney.

All the remaining portion of the total energy expended—the useful part—is transmitted to some other area, where it, too, ultimately returns to the environment as thermal pollution after performing its duty in a television set, refrigerator, power saw, electric train, or light bulb.

The local waste heat at the site of a power plant may be considered "good" or "bad," depending upon the circumstances and the cleverness with which it is used. For example, it is planned to employ part of the waste heat from a projected nuclear plant to help hatch salmon eggs for later release into the Skagit River in Washington State.[6] In Sweden, it has been proposed to place nuclear power plants near main cities and use the waste heat for home and other heating purposes.[7]

Although it has not been suggested before to our knowledge, a sensible use might be to warm the waters of a cove off such a body of water as Puget Sound in the Pacific Northwest to a reasonable swimming temperature, thus creating a salt-water beach resort of quality unparalleled within a radius of 1,000 miles. The proposal is not frivolous; the radioactivity of water discharged from a nuclear plant is less than $1/35$th the radioactivity of the natural isotopes in ocean water itself.[8]

Nuclear reactors of advanced types, such as the HTGR's, and breeders are expected to maintain efficiencies equal to those of the best coal-fired plants, about 40%.[9]

REFERENCES

1. In one type of reactor, the High Temperature Gas Cooled Reactor, the uranium is about 90 percent U-235, but mixed with Th-232.

2. William Brobst, *Transportation of Nuclear Fuel and Waste.* Nuclear Technology, vol. *24*, p. 343 (Dec. 1974).

3. H. York, *Race to Oblivion,* Simon and Schuster (1971), p. 61-68.

4. R. G. Hewlett and F. Duncan, *Atomic Shield,* page 497, Penn. State University Press (1972). Also republished by USAEC (WASH 1214, Vol. II) (1972.).

    Alvin M. Weinberg, *The Maturity and Future of Nuclear Energy,* American Scientist, vol. 64, p. 16 (Jan.-Feb. 1976).

5. *Nuclear Power and the Environment,* American Nuclear Society, p. 17 (March, 1974).

6. H. Williams, Seattle Times, Aug. 8, 1974.

7. See ref. 22, Chapter 13.

8. Ref. 5, p. 11.

9. T. H. Pigford, *Environmental Aspects of Nuclear Energy Production,* Annual. Rev. of Nuclear Science *24,* 515 (1974), p. 555.

COSTS AND USES OF ELECTRIC POWER
Chapter Eight

For more than a decade after fission reactors became technically feasible, they appeared to offer little economic advantage. By the 1960's, however, improvements in design and economies of scale made the situation sufficiently favorable for nuclear reactors for utilities to begin investing substantially in them. In consequence, by 1975 about 9 percent of the electricity used in the United States was nuclear generated.

It proved to have been a good investment, especially in view of the recent rises in fuel prices. According to a utility survey carried out by the Atomic Industrial Forum,[1] the average costs in 1975 of generating electrical energy were 12.3 mills per kilowatt-hour for nuclear sources, 17.5 mills per kilowatt-hour for coal, and 33.4 mills per kilowatt-hour for oil. These costs include both fuel expense and amortized capital costs, but do not include transmission and distribution costs.

It is seen that oil has been driven completely out of a competitive economic position by the price rises of the past few years, and coal, too, is substantially more expensive than nuclear. Although not included in this survey, hydroelectric power is the cheapest of all.

There have been numerous projections as to the future. All costs are going up, and utilities, faced with the choice of different directions in which to proceed for new facilities, are continually making and commissioning studies. Only two contenders appear in these studies: coal and nuclear. At any plausible price for oil, oil-fired plants are far too expensive, except for specialized places on a small scale.

Hydroelectric power is economically attractive, but the potentialities for expansion are extremely limited, even in the most favorable parts of the country. Solar generation of electricity is too remote for any very concrete estimates to have been made by utilities, but preliminary estimates indicate that it will be far more expensive than nuclear or coal. Supplies of natural gas are so limited that it is scarcely mentioned for future electric generating facilities.

In general, as one contemplates future electricity generating facilities, it appears that nuclear power continues to hold an edge over coal, but there is some uncertainty as to how decisive, or universal, this edge is. In some localities, where coal with a low sulfur content is readily available, coal may prove to have an economic advantage. Under other conditions, studies show a big margin for nuclear energy. For example, a study made by a Kansas utility concludes that it would possibly save a billion or more dollars over a 30 year period by proceeding with a nuclear plant, in preference to a coal plant.[2]

A recent study has been made by the municipal electric company of Seattle, Seattle City Light.[3] It compares costs (in 1975 dollars) for coal and nuclear plants, each generating 1,000 megawatts. The plants would be located in Western Washington. The projected capital cost of the nuclear plant is $590 million, while for the coal plant it is $460 million (assuming 80 percent SO_2 removal). Fuel costs are anticipated to be less than half as great for the nuclear reactor as for the coal plant. Considering all costs, including operating costs, the nuclear plant has an overall edge of about 5 or 10 percent over the coal plant, assuming both plants operate the same fraction of the time. Seattle is relatively close to sources of low sulfur coal; in the East, where coal costs are higher, the comparison is likely to come out more strongly in favor of nuclear energy.

This is borne out by data published by another utility, Commonwealth Edison of Chicago, which concludes that nuclear power has a large advantage over coal.[4] It cites capital costs of $421 per installed kilowatt for nuclear vs $420 for coal (with sulfur dioxide scrubbers). When fuel costs are taken into account, nuclear generated electricity would cost the consumer 24 mills per kilowatt-hour, and coal would cost 35 mills per kilowatt-hour. The figures are in dollars adjusted to 1974 price levels, but refer to power plants whose construction was started in 1975. The figures take an added significance because Commonwealth Edison is already generating about 35 percent of its electricity by nuclear power, and is therefore highly experienced.

In addition, nuclear plants have a further important advantage. Once built, a nuclear power plant is relatively invulnerable to continued inflation. Its chief costs after construction consist of paying off, at a fixed rate, the capital construction debt; increases in fuel price levels will have little effect on the cost of electricity.

On the other hand, oil- and coal-fired plants are immediately vulnerable to fuel-price changes. The worldwide story of sudden oil price rises is well known. Less well known is the fact that some utilities paid as much as four times more for a ton of coal in September, 1974, than they did in 1973. But these are exceptional cases; on the average coal increased in price from $8.50 per ton in 1973 to $15 per ton in 1974.[5]

The chief difference between future prices of oil and of coal is that oil prices can be imposed by an international cartel over which the U.S. has essentially no control, whereas domestic, economic and political forces control the price of (domestic) coal.

Nuclear power and hydropower share an advantage which derives from low fuel, or direct, costs: Once a nuclear plant, or a hydropower dam, is built, the cost of electricity is nearly independent of total output. (Of course, this conclusion applies to hydropower only if abundant water is available, a condition not always met, even in the water-rich Pacific Northwest.)

It follows that production of electricity in off-peak hours adds very little to total operating costs. The advantage, already considerable for present-day light water reactors, will be even greater for breeder reactors, fuel costs for which are expected to be one-tenth (or less) as much.

Since the peak demand for electricity is often twice the minimum—and the peak is equivalent to the plant's capacity—the normal power demand in off-peak hours can be as little as a third of the total possible energy output. As a result, electric power can be put to use in ways which might otherwise be uneconomical.

Typical variation of demand for electricity through the day. Usually, the demand is highest in early evening, and more power, i.e. "peaking power", is needed then.

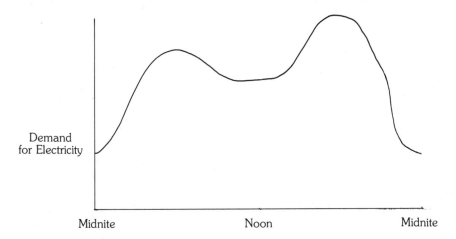

In the Northwest this is already done by manufacturing aluminum—a process requiring large electrical energy use—during off-peak hours. Were this not done, much water would simply "go over the dam" unused.

Another practice followed in the Pacific Northwest is to send power to California, thus enabling Southern California oil-fired plants to reduce their output. Thus, water is being poured on the troubled oil scene via the intermediary of electricity.

There are other proposals to use the off-peak power of nuclear plants. One is to produce hydrogen gas by electrolysis of water. Schemes are under study to utilize the hydrogen directly as a fuel, or to combine the hydrogen with carbon dioxide from the air to make a very clean fuel for automobiles or other uses.

Another possibility is to manufacture fertilizer. Indeed, what better use could be found for using the off-peak hours of breeder-reactor power plants than effectively to convert otherwise useless uranium into food?

It is interesting to compare various energy sources by reducing the basic cost of each to the same energy unit—the kilowatt-hour. In these comparisons the cost of conversion to a useful form, such as mechanical work or heat, is not included. The energy chart in Appendix A is a convenient aid for these comparisons. In these calculations, the full energy content of fuel is considered and no allowance is made for losses, such as the loss of about two-thirds of the energy in converting to electricity in a power plant.

Food for people is, of course, one of our most expensive forms of energy. The reasons are discussed in Chapter 2. Food costs about 55 cents (550 mills) per kilowatt-hour, on the average, for the average American diet[6] as of 1975.

Costs of other common fuels, when compared in this way, are: Gasoline at 55 cents per gallon is equivalent to 12.5 mills per kilowatt-hour; $25 per ton coal—3.3 mills per kilowatt-hour; and uranium at $30 per pound—3.5 thousandths of a mill per kilowatt-hour with breeder reactors and about 0.5 mills per kilowatt-hour with light water reactors.

Another comparison is the cost of electrical energy as supplied to a homeowner. This ranges from a low of about 10 mills per kilowatt-hour (Seattle) to perhaps 60 or 70 mills per kilowatt-hour in other regions.

It is interesting to note that when the efficiency of conversion of a typical home-heating oil burner is taken into account (about 60 percent), it is sometimes cheaper to heat a home even now with an electric furnace than with oil. And with heat pumps, which have efficiencies of 200 or more percent (as described in Chapter 4), the "breakeven-point" between fuel oil at 37 cents per gallon and electricity is about 30 mills per kilowatt-hour for the delivered electrical energy.

Since there is little likelihood that prices of fuel oil will fall in the future, home-heating with coal or nuclear-generated electricity utilizing heat pumps can be economically advantageous, as well as being highly desirable from the standpoint of conserving oil.

Although coal was once widely used for direct home heating, there has been no serious suggestion that, despite the energy problem, we return to this method today. The reason is that small coal burning furnaces incorporating adequate devices for removal of pollutants are not feasible.

In the U.S., ever since about 1910, the consumption of electrical energy has increased at a rate about two and a half times faster than consumption of all forms of energy—7.0 percent per year, vs 2.7 percent per year, respectively. A portion, about 1.8 percent per year, of these increases has been due to the increase in population.

The reasons for the increasing reliance on the use of electricity are clear: It is exceedingly useful, convenient, and clean for the consumer. It can be substituted in many instances in which we depend on oil. We can run trains with it. We can operate urban transportation systems with it. And we have no alternative in applications such as communications, motive power to operate oil burners, dishwashers, and washing machines, and most important of all, for lighting of our homes and streets. How unsafe would our streets be without electric lights!

A number of factors suggest that society will move more and more towards increased use of electricity: the dwindling reserves of oil and gas on a world wide scale; the difficulty in using coal for anything other than production of electricity and industrial process heat; and the virtual impossibility (at present) of using nuclear energy except for electricity generation and ship propulsion. Fortunately much of the technology needed for increased electricity utilization is either already at hand or proven technically feasible; it remains to verify that some new proposals, such as hydrogen production, are economically viable.

In summary, the U.S. economy could eventually be maintained almost completely on an electrical basis. But this can be accomplished only with a good deal of advance planning. The planning will not take place unless it can be assumed on a realistic basis that an adequate electric-power generation capacity will exist in the future.

REFERENCES

1. *Nuclear Info*, Atomic Industrial Forum, March, 1976.

2. J. R. Lucas, *KG&E Cost and Savings Comparison with Sensitivity Analysis*, Kansas Gas and Electric Company, June, 1975.

3. *Energy 1990 Study, Initial Report*, Volume 1, Chapter 4, Seattle City Light, February, 1976.

4. Nucleonics Week 17, January 8, 1976.

5. R. Gillette, Science *187*, 43 (January 10, 1975).

6. The price of the average "market basket" for the average American family of 3.2 persons is about $1,800 a year. A family eats a total of about 10,000 calories a day. The average cost per calorie can then be calculated and the result converted into any other energy unit. It calculates out to about 55 cents per kilowatt-hour of food energy. Food, of course, also contains other essential ingredients such as protein and vitamins, which are included in the above cost.

WHY THE DISPUTE?
Chapter Nine

Facts and figures offered thus far would lead to the conclusion, it seems, that the advantages of nuclear power are so overwhelming society might be expected to accept it as an unmitigated blessing and look upon those who helped create it as saviors of mankind.

Judging from public criticism of nuclear power, however, that is far from the general reaction. What is wrong with nuclear power, then? Is there anything *really* wrong? Are the drawbacks of nuclear power real or imaginary? Do we have any genuine alternatives?

In trying to find the answers, we should begin by trying to clear up some simple misconceptions. The other, more serious issues will be treated in later chapters.

Thermal pollution has often been believed mistakenly to be an evil unique to nuclear-power plants. As indicated in an earlier chapter, that is simply not true. Although present-day light water reactors (LWR's) produce more thermal pollution than the latest coal-fired plants, the difference is small, and future nuclear plants, the breeder reactors, are expected to produce the same amount of thermal waste as late-model coal-fired plants.

Opponents of nuclear power have insisted nuclear power is more expensive than fossil-fueled power and other forms. The fact is only hydroelectric power is cheaper than nuclear power, and the number of dams that can be built to produce hydro power has almost reached its limit already.

Even if one chooses to disbelieve the figures on power costs quoted in Chapter 8, it is hard to argue away the simple fact that many electric-power utilities are ordering nuclear-powered facilities. Utilities companies would hardly be expected to pursue capriciously a *more* expensive course, particularly in view of the obstacles they must overcome in combating the "environmentalist" challenges each new reactor now faces.

51

Some arguments against nuclear power have been couched in terms of energy input and output. Perhaps the most striking of the criticisms are the assertions that nuclear reactors require almost as much energy for their construction and operation as they are expected to produce in their projected lifetimes. This is a groundless concern. Actually, in an assumed life of 30 years, a reactor can be expected to produce roughly 15 times the energy required to build and operate it.[1,2] The precise number depends upon the quality of the uranium ore used, the type of reactor, the reliability assumed for the reactor, and the allowance made for the fact that the electrical energy produced by a power plant is more valuable than the thermal energy which has been consumed in its construction.

However, despite all these variables, the factor of 15 is a reasonably reliable estimate, because by far the largest part of the energy input is the energy required to convert natural uranium (0.7 percent U-235) to enriched uranium (3 percent U-235). This requires close to 5 percent as much energy as is obtained from running the reactor. Other energy costs, such as those for constructing the reactor, add to the energy input, bringing the output-to-input ratio down from 20, for uranium enrichment alone, to about 15.

Another number which is sometimes cited is the payback time—the time required for the reactor to produce as much energy as was required to build it. Some industry estimates[3] report that the initial investment is recouped within several months. One can calculate a payback time of up to one year if one includes, along with the reactor itself, the energy costs of the initial (three-year) fuel supply and the reactor's share of the electricity transmission and distribution network.

But no matter what criteria are used, each reactor in its 30-year lifetime will pay for itself many times over, and the payback of the initial energy input comes within a short time after the reactor goes into operation.

A more sophisticated argument, however, is sometimes made to the effect that a rapidly expanding nuclear program will not pay for itself as a whole—or at least that the point of paying itself off will not be reached for a very long time.[4,5] The rationale is that for rapid exponential expansion, the numerous plants under construction consume energy in amounts comparable to the energy produced by the relatively small number of plants in operation. With exponential growth, the argument goes, the number of plants under construction keeps increasing, causing great difficulty in catching up.

While mathematically valid, the argument is relevant only in an imaginary world in which our choices would be between never-ending growth and no-growth.

More realistically,[6] if initial growth is rapid, it will soon saturate reasonable demand; as soon as the growth tapers off, one begins to recoup the investment dramatically. Obviously, any new facilities require an investment in money and energy. Nuclear plants are not qualitatively unique in this respect. They have somewhat higher capital costs than other power plants, but that is offset by the fact that nuclear plants make possible a saving in fossil-fuel energy by lessening our reliance on oil, gas, or coal-fired power plants. The important fossil-fuel saving is the crucial consideration in a rational energy-conservation program, in view of the scarcity of fossil fuels, their usefulness to the chemical industry, and their major role in meeting energy requirements other than for electric-power generation.

In summary, if a person wished to save energy in the short run, he should not build new plants of any sort, including nuclear reactors. However, if he wishes to conserve fossil fuels for the long run, investment in nuclear-power plants is advisable. As expressed by Gerald Leach,[6] ". . . nuclear systems are some 3 to 80 times better at converting fossil fuels to electricity than are today's conventional power stations." (This assumes fossil-fuel energy is expended during the construction of nuclear-power plants.) Sensible energy analysis can shed light on the details of the comparison between short-term costs and long-term benefits, but in any case, the existence of the long-term benefits is clear.

Still another controversy concerns the reliability of existing reactors. It is contended they are simply not working very well. Unquestionably, some have been plagued by technical troubles, particularly in the time of early operation. Many of the troubles are not associated with the nuclear aspects of the plants and would be common to coal-fired or any other type of large industrial plants. Nuclear plants, however, are undoubtedly more complicated and quite new technologically; the "growing pains" are not unexpected.

There are two common ways of measuring the overall reliability of power plants: the "availability factor" and the "capacity factor." The availability factor gives the fraction of the time the plant was available for use, had its power been needed. The capacity factor is a more comprehensive figure, giving the actual output of the plant, as compared to the rated output if it ran at full capacity 24 hours a day, 365 days a year. Clearly, the availability factors will always be larger than the capacity factors, because sometimes the power is not needed and sometimes the plant may be operating at less than its peak output.

A comparison was made by Commonwealth Edison between its nuclear plants and its newest coal plants for the year ending August 31, 1974.[7] Commonwealth Edison operates five large nuclear plants in the Chicago area, making that region the heaviest user of nuclear power in the country. For the period surveyed, the nuclear plants were available 75.5 percent of the time, whereas the coal-fired plants were available only 67.6 percent of the time—a sizable difference, in this case, in favor of nuclear.

More recent and complete figures for nuclear plants can be obtained from tabulations of the performance of each reactor in the country.[8] Leaving out those plants which just started up in 1975, the overall capacity factor for U.S. nuclear reactors in 1975 was 60 percent, slightly better than the comparable figure for 1974, but still short of a reasonable goal of 80 percent. One never can hope to achieve 100 percent, because the reactor must be shut down for refueling, if for no other reason. In addition, the tight restrictions of the Nuclear Regulatory Commission will force a plant to shut down for inspection and possible modifications not only when there is a small hint of trouble in it, but also when there is a hint of trouble in another plant of similar design.

Although the capacity factor for nuclear plants is not as high as desired and presumably as will ultimately be achieved, it was higher than that for coal-fired plants in 1975 and much higher than for oil-fired plants. This does not mean the nuclear plants as a whole have already achieved a higher level of reliability than fossil-fueled plants. Their *availability* factors were lower than those for either coal or oil plants, despite their higher *capacity* factors.[9] But the low fuel cost makes the nuclear plants more economical than the fossil fuel plants, and hence they are used when there is a choice.

Thus, as far as "running smoothly" is concerned, we see a fairly good and improving record for nuclear plants. Nuclear reactors have clearly proved they do their job most of the time, enough of the time to make them economically attractive. It remains to improve on the "most" without sacrificing the present high standards for safety in operation.

There are five more controversial issues associated with nuclear power, only one of which is relevant also for coal-fired plants. These are the real causes of the fight over nuclear power. They are:

1. Radiation emission hazards in normal operation. Coal-fired plants also produce a similar hazard, a fact seldom appreciated.

2. Radiation hazards in case of abnormal operation caused by some internal accident or component failure. In this category we can also include the possibility of radiation hazard produced by sabotage, earthquakes, war, or other externally induced accident. For short, this issue is referred to as Reactor Safety.

3. The disposal of the radioactive nuclear wastes.

4. The possibility of diversion, or theft, of nuclear fuels by individuals or groups for the purpose of making, or threatening to make, a bomb, or of producing widespread radioactive contamination.

5. International proliferation of nuclear power plants and nuclear weapons.

Each of these issues is discussed at length separately in the chapters which follow.

REFERENCES

1. C. T. Rombaugh and B. V. Koen, *Total Energy Investment in Nuclear Power Plants,* Nuclear Technology *20,* 5 (1975).

2. F. von Hippel, M. Fels, and H. Krugman, *The Net Energy from Nuclear Reactors,* FAS Bulletin *3,* 5 (April, 1975).

3. Kenneth Davis, as cited in Nucleonics Week, p. 7, March 6, 1975.

4. P. Chapman, *The Ins and Outs of Nuclear Power,* New Scientist *64,* 866 (Dec. 19, 1974).

5. J. Price, *Dynamic Energy Analysis and Nuclear Power,* December, 1974.

6. G. Leach, New Scientist *65,* 60 (Jan. 16, 1975).

7. Nucleonics Week, Sept. 19, 1974.

8. Ibid, Jan. 29, 1976.

9. Based on summary for first nine months of 1975 in *Nuclear Info,* Atomic Industrial Forum, January, 1976.

RADIATION HAZARDS
Chapter Ten

In normal operation a minute quantity of radioactivity escapes from nuclear-power plants. The amount released from the most recently designed plants is less than from those built earlier. For all power plants, however, the incremental radiation even at the boundary of the plant site is far below natural background radiation, and most experts agree the radiation produced by nuclear plants constitutes a very minor risk to the public.

On the other hand, it is prudent to assume that any amount of radiation, however small, might be harmful to some degree. It is the awareness of this possibility, often divorced from any realistic consideration of the magnitudes of the risks, that provides the basis for some of the existing fears of nuclear reactors. Contributing to these fears is the vague feeling that radiation represents a new and strange hazard, and therefore that we should be doubly wary of it.

The feeling of strangeness is based on a profound misconception. In fact, the situation is just the reverse. Not only has radiation been extensively studied in the laboratory over the past 50 years, but the human race, over millions of years, has gained far more experience with nuclear radiation than with almost any other environmental hazard. Each star is a giant nuclear reactor, and the solar system itself was formed some five billion years ago from the debris of past nuclear reactions and explosions. Despite the long lapse of time, many lingering radioactive species remain on earth and produce a continual radiation exposure for everyone alive. Even within our bodies (as well as those of our ancestors) there is an appreciable amount of radioactivity mostly in the form of the isotope potassium-40, which provides a constant internal source of radiation. In addition, continuing violent cosmic activity results in the bombarding of the earth's atmosphere with rapidly moving atomic nuclei, the so-called cosmic rays. These cosmic rays produce a radiation exposure essentially identical to that produced by radioactive emissions. Thus, all life has developed in—and continues to exist in—an environment which guarantees a continual exposure to radiation.

Therefore, human experience demonstrates a certain amount of radiation is not destructive, at least in an over-all historical sense. But is it good or bad?

From the standpoint of the human race, it can be conjectured that over millions of years the net effect of natural radiation has been beneficial in that the mutations radiation may have caused have led to the development of the species in generally favorable directions. From the standpoint of any one individual, however, radiation is probably harmful in that it could enhance the chance of cancer or of genetic damage to one's offspring. Furthermore, in considering nuclear reactors, we are talking about increments above the natural radiation level or above any other sources individuals now accept— and hence, both from the individual and ecological standpoints, any significant increases should be viewed critically.

One way to estimate whether a given incremental amount of radiation is significant is to compare it with the natural level of radiation or, perhaps even more meaningfully, to variations in the natural level of radiation. For the U.S. as a whole, the average person receives a radiation dose from natural sources of about 130 mrem per year.[1] The dosage unit used here is the millirem (abbreviated mrem) per year. It is one-thousandth of a "Roentgen equivalent man." In Colorado, however, the average person's dose is 250 mrem per year because of the higher altitude, which increases the cosmic ray intensity, and because of the minerals in the ground, which contain more than an average amount of radioactivity.

In Louisiana, on the other hand, the average dose is only about 100 mrem per year. No evidence exists that indicates the people in Denver suffer excess health damage from radiation, nor has any serious suggestion been made that people should move from Denver to New Orleans to escape the added radiation. It appears, then, that a general consensus prevails among experts and the public alike that a dose of 100 mrem per year of radiation is not so harmful as to induce us to make major changes in our lifestyle to avoid it.

Another measure of how seriously we as a society view radiation is the public's attitude in accepting medical X-ray exposures as worthwhile. The average abdominal dose per capita in the entire U.S. in 1970 was estimated to be 72 mrem per year,[1] with some individuals far exceeding that amount. While efforts are under way to reduce what has often been careless or excessive medical radiation exposure, it is evident once again that society, including medical science, is somewhat undisturbed about doses in the neighborhood of 100 mrem. The prudence of such a casual attitude has been the subject of substantial criticism in recent times, and the general feeling of the scientific community and governmental agencies is that any gratuitous, significant radiation is undesirable, even though there is no *positive* evidence that it is harmful.

According to the best available estimates,[2] a 100-mrem increment in radiation might give an individual a small extra likelihood of contracting cancer or leukemia, estimated to be about one chance in 50,000. (The reason the word "might" is used is amplified in Appendix C.) For one individual, then, the effect could seem minor, but if projected throughout the entire population, it could entail a very serious human loss.

Whatever the implications of such concerns for medical X-rays, however, they have virtually nothing to do with the dangers from nuclear-power plants in normal operation, because the plants produce much less than 100 mrem per year of external radiation.

Present federal regulations and actual practice limit the maximum radiation at the boundary of a nuclear-power plant to the equivalent total dose of about 10 mrem per year, allowing for different sources of radiation.[3] Thus, the incremental radiation received by a person residing continuously 24 hours a day at the fence of a nuclear reactor would not exceed $1/10$th of the natural environmental dose. It is less, for example, than the extra dose received if one lives in a brick house, with its radioactive minerals, rather than in a wood house.[4]

Moreover, the amount of radiation a person could receive decreases rapidly as he moves away from the reactor boundary. In consequence, the U.S. Environmental Protection Agency[1] estimates that the radiation dose for the average person, assuming a rapidly expanding nuclear program, will be well below 0.1 mrem per year in 1980 and less than 0.5 mrem per year by the year 2000. This includes the radiation from all operations in the fuel cycle, including the reactors themselves and the plants for reprocessing the fuel. An exposure of 0.5 mrem is less than the additional radiation one receives by taking a weekend trip in Colorado, or by living one year in a house 100 feet higher than one's old house.

Such small doses from nuclear power cannot be taken very seriously, and attempts to put them in perspective often appear flippant. One of the more striking comparisons has been related by Dr. Alvin Weinberg,[5] formerly director of energy research and development in the Federal Energy Office. While he was head of the Oak Ridge National Laboratory, he had an estimate made of the *radiation* one receives from another person sleeping in the same bed. The radiation involved comes from the internal potassium-40 that is naturally present in every human being. As indicated already, potassium-40 has nothing to do with nuclear power; it is part of our primal heritage from the universe. The dose we receive from our own internal radioactivity is about 20 mrem per year, and the additional radiation dose drawn from a bed partner, presumably assuming average sleeping habits, was reported by Dr. Weinberg to have been estimated at 0.3 mrem per year.

Thus, these somewhat whimsically developed figures indicate the average married couple would at present gain far more in terms of decreased radiation exposure by sleeping in separate beds than by shutting down the entire civilian nuclear-power program. Which should they choose to give up, their natural warmth or their industrial warmth? The sensible answer, of course, is neither.

In the light of such numbers and the overall evidence, it has become generally agreed, even by most critics of nuclear power, that the environmental contamination from nuclear reactors *in normal operation* is not a matter of serious concern. The nuclear reactor can be viewed as a benign neighbor, compared to coal-fired plants which emit radioactive materials in amounts comparable to nuclear power plants, as well as chemical pollutants which have much more significant health effects.

In addition to the reactors themselves, another source of environmental contamination from normal operation of the total nuclear power cycle lies in possible radioactive releases from future fuel reprocessing plants. However, stringent limitations on radiation levels apply to these facilities as well, and their contribution is included in the encouraging Environmental Protection Agency estimates of total radiation dose, mentioned above.

REFERENCES

1. *Estimates of Ionizing Radiation Doses in the United States 1960-2000,* U.S. Environmental Protection Agency, ORP/CSD 72-1 (1972).

2. *The Effects on Populations of Exposure to Low Levels of Ionizing Radiations* ("BEIR Report"), National Academy of Sciences-National Research Council, (November, 1972).

3. These restrictions are embodied in the federal standards on *Licensing of Production and Utilization Facilities,* 10CFR50, Appendix I (1975).

4. *Nuclear Power and the Environment,* American Nuclear Society (March, 1974), p. 9.

5. A. Weinberg, International Conference on Radiology and Radiation Biology, Seattle, Washington (July 15, 1974).

PLUTONIUM AND HUMAN HEALTH
Chapter Eleven

The element plutonium has figured widely in nuclear debates since 1974, and seems destined to remain in the public eye for some time. In some ways it is an unusual element, in that virtually no plutonium exists naturally on earth. All the isotopes of plutonium are unstable, and in the billions of years since the earth's elements were made the original plutonium has long since decayed away.

However, although there is no "natural" plutonium, plutonium is manufactured in uranium fueled reactors. A large light water reactor produces about 500 pounds of plutonium per year. If extracted from the used fuel and returned to another reactor, the plutonium can substitute for about one-fourth the uranium otherwise needed, and thus is a useful product. Plutonium's potential in breeder reactors may turn out to make it the major fuel of the future.

But it is also a troublesome product, partly because it is highly toxic and partly because it could possibly be used to make bombs. Problems concerning disposal of plutonium waste will be considered in Chapter 12, and the bomb issue will be discussed in Chapter 15; in this chapter we concentrate on the toxicity of plutonium.

Plutonium has received spectacular and hostile attention in the press. An article about it[1] has been entitled "The Element of the Lord of Hell?" It has been characterized as being "perhaps the most toxic substance known."[2] A college newspaper[3] has quoted testimony that "One pound of plutonium, ground finely, and dispersed properly in the earth's atmosphere, would cause lung cancer in everyone on earth." These statements, in themselves, have more to do with mythology, ancient and modern, than with science. Gram for gram, plutonium is less toxic than botulism toxin or tetanus toxin, and the last quotation exaggerates the danger about one-hundred-million-fold.

But it is not a myth that plutonium is a highly toxic substance. Whether, microgram for microgram, it is more or less toxic than some of the exotic biological and chemical alternatives is not the key to judging the dangers. The real issue is the toxicity of plutonium itself, coupled with gauging the possiblity that people will be exposed to it.

Study of the medical effects of plutonium dates back to 1944, when it was first being produced in the atomic bomb program of World War II. Very large programs to study its toxicity have been underway for over 20 years, and much is known about the passage of plutonium through the body, and the chief sources of danger.

For example, plutonium taken into the body with food or drink is retained only to the extent of one part per million for insoluble compounds and 30 parts per million for soluble compounds.[4] In consequence, ingestion of plutonium is not strikingly hazardous. It is not very much worse than ordinary lead for ingestion.

Inhalation of plutonium, on the other hand, is much more dangerous, because a far higher proportion of the plutonium is retained in the body. Concern over plutonium toxicity has thus centered upon the possible consequences of breathing it. Safety standards, setting limits on the intake of plutonium by plutonium workers are more than 1,000 times stricter for inhalation than for ingestion. There are no known cases of cancer from plutonium in man, so these standards are based on observed effects of radium in man and on comparisons of radium and plutonium in animals. The Linearity Hypothesis (Appendix C) is assumed, and in that sense, as discussed in Chapter 10, the standards are conservative ones.

However, a theory put forth recently suggested that the true radiotoxicity of plutonium for producing cancer is about one-hundred thousand times greater than has been supposed.[5] It is called the "hot-particle" theory and is based on the idea that plutonium can lodge in small concentrated aggregates in particularly sensitive portions of the lung, where it is then claimed to be very effective in producing cancers. This general idea has been considered since as early as 1950, but it had never been widely believed. In view of the importance of the problem, and the vigor with which the hot-particle hypothesis has been recently advanced, especially in writings by Dr. Arthur Tamplin and Dr. Thomas Cochran, the issue warranted a new careful study, and, if possible, a verification or refutation of the hot-particle hypothesis.

Two reviews of the entire matter of the toxicity of plutonium have been made by highly regarded specialists. One was carried out by a committee of the British Medical Research Council.[6] The report of this study[7] concludes, "there is no evidence that irradiation by 'hot particles' in the lung is markedly more hazardous than the same activity uniformly distributed or that the currently recommended standards for inhalation of plutonium are seriously in error." The chairman of the committee preparing the report, Dr. Robin Mole, later stated his evaluation of the papers of Tamplin and Cochran on the hot-particle hypothesis in blunter terms:[8] "It really was a penance for me to have to read such ill-written and ill-argued documents. Why should attention be paid to inadequate science?"

The second study[9] was carried out by a group of American experts. Their conclusions are similar to those in the British report. They found that neither "consideration of mechanisms of radiation carcinogenesis" nor "empirical considerations" lend support to the hot-particle hypothesis. Although they did not indulge in the ascerbic rhetoric of Mole, their rejection of the hot-particle hypothesis is as definite.

Conclusions reached by the two studies are in such great contrast to many popular statements that it is of interest to inquire about the evidence on which the British and U.S. arguments were based. Some of this argument is of a rather technical nature and will not be repeated here, but in addition, very simple deductions can be made from the results of two series of accidental exposures in which plutonium was inhaled by workers.

In the Second World War, 25 persons accidentally acquired body burdens of plutonium at the bomb laboratory in Los Alamos. It is known that some of that plutonium still resides in their bodies in amounts substantially in excess of the figure the hot-particle theory indicates would produce cancer. The 25 have been studied constantly and with great care—as can be imagined—ever since, or for a period of about 30 years. Had the original hot-particle hypothesis been correct, it is estimated that these men would together have received doses sufficient to produce about 5,000 cancer tumors in their lungs.[10] In fact, as of 1974, there has been no incidence of lung cancer among them.

Along similar lines, a fire in a plutonium fabrication plant at Rocky Flats, Colorado, in 1965 resulted in the heavy exposure of 25 people. According to the original hot-particle hypothesis, they should eventually accumulate 5 to 50 cancers *each*.[11] None has shown any sign of cancer tumors, although by now some cancers would have been expected.

In the face of such evidence, Tamplin and Cochran have revised upward their estimates of the size of the hot particle needed to produce cancer.[1] Even earlier, Tamplin had reduced the claimed error in the standards from 115,000 to 1,000.[12] But it is unlikely that their new estimates will receive any greater acceptance than their earlier ones.

The apparent demise of the hot-particle hypothesis does not promise to end the controversy over plutonium toxicity. Already, new suggestions have been made that it is still much more toxic than believed by the majority of experts. One would hope that the hot-particle experience will contribute to a more skeptical, although not totally closed-minded, response to these further allegations.

Perhaps the greatest reason for concern about plutonium toxicity is the possibility that it will be intentionally dispersed by a terrorist group. What then will happen? The most reasonable guide to the toxicity is provided by the estimates of the International Council for Radiation Protection. In a study of the terrorist problem,[13] Professor Bernard Cohen adopted these, and also made estimates of the extent to which plutonium dispersed by terrorists might actually be ingested or inhaled. His analysis indicates that there would be one cancer death per 15 grams of plutonium dispersed, or 30 per pound. This is to be contrasted to the statement quoted earlier, where it was said that one pound could give lung cancer to everyone on earth—about 4 billion people.

The difference between these numbers depends partly on the acceptance or rejection of the hot-particle hypothesis. It also depends on what one assumes happens to the plutonium after it is dispersed. Cohen's calculations include realistic consideration of how dust particles spread through the atmosphere and the rate at which they can be taken in by breathing. The higher number requires not only the hot-particle hypothesis, but also inhalation of almost every speck of plutonium by one individual or another (in specks of just the right size), a nonsensical premise.

In his calculation, Cohen assumed the plutonium to be dispersed without warning in a city with a somewhat typical population density of 10,000 people per square kilometer. With warning, and accompanying simple protective measures, this cancer rate could be significantly reduced. Contamination of a city's water supply with plutonium, this same study concludes, is even less effective, requiring over 10 pounds to produce one cancer death. It should be noted that these deaths are by no means certain, in that the linearity hypothesis may have caused an over-estimate of the true effects. Further, the deaths occur from cancers, typically one or more decades after exposure, and thus would not provide the quick dramatic impact desired by terrorists, except perhaps through the fear engendered. Other aspects of plutonium as a possible terrorist tool will be considered in later chapters.

In the preceding discussion we have criticized what we believe to be wild exaggerations of the extent of plutonium toxicity. However, it is important to recognize that, by any reckoning, plutonium is highly hazardous. All evidence shows that very small amounts pose non-negligible cancer risks if inhaled into the lung. In view of the amounts of plutonium to be produced in the future continued great care must be exercised in its handling. We can gain confidence in our ability to handle it with the requisite care from the fact that despite the handling of many tens of tons of plutonium in the weapons programs, there is no known incidence of plutonium induced cancer in weapons industry workers, or, for that matter, in anyone else.[14]

REFERENCES

1. Arthur Tamplin and Thomas Cochran, *The Element of the Lord of Hell?*, New Scientist (May 29, 1975).

2. J. G. Speth, A. R. Tamplin, and T. B. Cochran, *The Plutonium Decision*, National Resources Defense Council, Washington, D.C. (September, 1974) p. 1.

3. University of Washington Daily, Seattle, Washington (Oct. 23, 1975).

4. Ref. 6, pages 26-27.

5. Arthur Tamplin and Thomas Cochran, *Radiation Standards for Hot Particles*, Natural Resources Defense Council, Washington, D.C. (February, 1974).

6. *The Toxicity of Plutonium,* Her Majesty's Stationary Office, London (1975).

7. Ibid., page 5.

8. Robin Mole, *Anxieties about Safety Standards*, New Scientist (May 29, 1975), page 506.

9. W. J. Bair, C. R. Richmond, and B. W. Wachholz, *A Radiobiological Assessment of the Spatial Distribution of Radiation Dose from Inhaled Plutonium*, WASH-1320, U.S. Atomic Energy Commission (September, 1974).

10. Ibid., page 26.

11. Ibid., page 28.

12. Arthur Tamplin, New Scientist (March 20, 1975).

13. Bernard L. Cohen, *The Hazards in Plutonium Dispersal*, (July, 1975).

14. Ref. 9, p. 25.

HANDLING THE WASTES
Chapter Twelve

T_o some persons, problems associated with disposal of the radioactive wastes produced by nuclear reactors have seemed to present major, even decisive, objections to the nuclear-power program. It is contended the U.S. is not only running immediate risks but is also imposing a continued dangerous burden on future generations. Typical of this concern is the following statement from Allen V. Kneese, director of the Resources for the Future program of studies in the quality of the environment:[1]

> "Here we are speaking of hazards that may affect humanity many generations hence and equity questions that can neither be neglected as inconsequential or evaluated on any known theoretical or empirical basis. This means that technical people, be they physicists or economists, cannot legitimately make the decision to generate such hazards. Our society confronts a problem of great moral profundity; in my opinion, it is one of the most consequential that has ever faced mankind. In a democratic society the only legitimate means for making such a choice is through the mechanisms of representative government."

No one can quarrel with the spirit of moral responsibility that produced Kneese's statement and others like it. However, if the people and government are to make reasoned decisions, technical information is necessary, including, in particular, a balanced perspective on the magnitude of the hazards under consideration.

To begin a review of the technical aspects of the problem, an important feature of radioactive substances must first be understood. Each radioactive species dies out with a characteristic time scale, called its half-life. After one half-life, half of the nuclei have decayed and the residue is only half as radioactive. After ten half-lives, only about 0.1 percent of the activity remains. After 20 half-lives, only about 1 part in a million remains.

Thus, all radioactive substances start "hot" and end "cold." The half-lives of different species vary enormously, from fractions of a second for some to billions of years for others.

For equal amounts of material present initially, the level of activity is greatest for the short half-lives, and it is lowest for the long half-lives. From the standpoint of waste disposal, the shortest half-lives are unimportant because the material decays away very quickly, and the longest half-lives are unimportant because the material is always relatively "cool." It is the intermediate half-lives that create a disposal hazard, and these range from years to thousands of years. Natural uranium is itself radioactive, but its half-life is so great that even large quantities are relatively innocuous. An example of an isotope at the lower end of the intermediate range is strontium-90 (29 years) and at the higher end of the range plutonium-239 (24,000 years).

The radioactive species in the waste happen to be neatly divided into two groups: the so-called actinides and the fission products. Actinides are heavy nuclei produced by the addition of neutrons to the uranium fuel and include, among other species, several isotopes of plutonium. In addition, they contain virtually all of the longer-lived activities and account for the storage problems remaining after 600 years.

On the other hand, the fission products are the result of the breakup of uranium (and to some extent plutonium), and are roughly half as heavy as uranium. They are responsible for almost all of the activity during the first few hundred years. The most important of these products from the standpoint of radiation hazard are the isotopes strontium-90 and cesium-137, with half-lives of 29 and 30 years, respectively. In 600 years their level of activity is reduced by about a million times, and neither they nor any of the other fission fragments (which by then are mostly stable isotopes) constitute an appreciable hazard.

When spent fuel is first removed from a reactor after about three years of service, it is permitted to cool for at least several months, during which time the most intense radioactivity dies away.[2] It is then ready for shipment to a reprocessing plant for the next step in waste handling. Shipment is in a very secure cask resembling a safe-deposit vault.[3]

The purpose of reprocessing is to turn the solid spent fuel into liquid and then to separate out chemically most of the uranium and all but about one-half percent of the plutonium. Plutonium extraction procedures were first developed in the 1940's for the nuclear weapons program, when the plutonium was used for bombs. However, extraction of plutonium and uranium also makes sense for the civilian power program, because the extracted material can be used as fuel in further reactors.

In addition, removing most of the plutonium makes the wastes less dangerous. It is the remaining ½ percent plutonium and the other actinides that constitute the source of the long-lived activity—and are the cause for much of the concern over waste-disposal.

One possible approach to long-term aspects of the problem is to perform a more complete chemical separation of the actinides, and then destroy them by returning them to a reactor. While no decision has been made to take this route, it is believed to be both technically and economically feasible.[4] On the other hand, recent studies indicate this to be unnecessary because, once 99.5 percent of the plutonium is removed, the remaining actinide activity is small enough to represent relatively little hazard.[5,6]

Although reprocessing of military wastes has been carried out for several decades, the reprocessing of commercial reactor fuel in the United States has been done only at a plant in New York which operated from 1966 to 1972. The facility has since been undergoing a major expansion, and it, and another reprocessing plant nearing completion in South Carolina, will not be ready to handle wastes for several years. In the meantime, the spent fuel is accumulating at the reactor sites, cooling down harmlessly, but standing as a symbol of poor planning. The program has been further delayed by a comprehensive reexamination of plutonium utilization and disposition issues being carried out by the Nuclear Regulatory Commission, a study stimulated in part by the alarm created by the flamboyant but apparently incorrect claims of the plutonium "hot-particle" advocates.

Once reprocessing begins again, the radioactive wastes, with whatever percentage of actinides remains, can be solidified into an impervious, glass-like material using one or another of several established processes.[7] In turn, the solid, glass-like material is to be encased in steel tubes or canisters, and then stored either in man-made surface mausoleums or in underground cavities.

It is relevant to stress that the volume of solid waste material to be dealt with is remarkably small. The volume of the solidified output of a large reactor running for a full year is about 70 cubic feet,[8] about the same as six standard filing cabinets. Assuming a rapidly expanding nuclear program, the total volume of waste generated annually in the United States in the year 2000 would be represented by a cube less than 40 feet on a side. Each person's lifetime share would be equivalent in size to about 75 aspirin tablets.

One typical plan calls for storing this waste in steel canisters, about 1 foot in diameter and 10 feet long. Ten such canisters would hold a year's waste from a single reactor. The canisters could be encased in concrete cylinders for further protection. After about ten years each canister would put out only about 3 kilowatts of heat,[8] as much as two or three bedroom electric heaters, and cooling them by natural air flow is not difficult.

For the next step, one possibility is to store the canisters and their containers in accessible, above-ground facilities. The area which would be required for this storage is very small. Slightly more than one square mile of desert land has been considered, perhaps at Hanford, Washington. At such a site, all wastes expected from an ambitious nuclear-power program up to or beyond the year 2000 could be accommodated.

It is planned that at some stage the wastes will be deposited in inaccessible and geologically stable underground sites. A transfer could be made from the temporary surface storage site after a few decades. Alternatively, the temporary surface storage step could be by-passed, and the waste canisters could be put underground sooner. In either case, salt mines are most often mentioned as likely underground candidates.

However, these waste handling plans have not found universal acceptance among critics in the general community and among some scientists. The principal objections are:

- Some hazards last for over 100,000 years, and no one can predict what will happen to the waste material over so long a time span.
- The AEC's advocacy of disposal in a Kansas salt mine was shown to be faulty, and thus the method is generally suspect.
- The AEC and its successors, ERDA and the NRC, still have not come up with a final plan for waste disposal.
- Large leaks of radioactive material at Hanford show the unreliability of the waste handling and monitoring procedures.
- Any storage facility is vulnerable to sabotage by terrorists.

The first concern has grown out of a recognition that plutonium-239 has a 24,000 year half-life. What has often been ignored is the fact that, on any reasonable basis of comparison, there is not much plutonium left in the wastes, assuming that 99.5 percent has been removed in reprocessing.

To appreciate this point one must have some picture of how the wastes could cause a problem. They will be deep underground, perhaps 2,000 feet below the surface. The only plausible way in which they could enter the biosphere would be if somehow water came into contact with them and carried radioactive material into our drinking water or into some part of our food chain.

A similar thing is already happening, because the natural radioactivity in the earth is being continually carried in small amounts into our rivers and into us. Therefore, it is interesting to compare the amount of natural activity in the earth with the amount which will be added from reactor waste. As an example, consider the wastes (all of them, not just plutonium) from 500 large reactors, each operating for 20 years. After the wastes sit for 600 years, the remaining activity from all these reactors will be less than 0.1 percent of the total natural activity in the earth's crust over the area of the United States down to a depth of 2,000 feet.[9] Thus, man's contribution to the earth's radioactivity will be small.

Of course, there are some differences between the wastes and the natural ores in the ground: the ores are in all sorts of locales, while the wastes will be in the safest and driest places we can find; the ores are rather spread out, while the wastes will be concentrated; and the ores are in a variety of physical forms, while the wastes will be glassified to be highly water resistant. The most surprising difference, however, is in the nature of the radioactive elements. The natural ores contain a good deal of radium, and, despite all the common impressions to the contrary, for a given amount of activity (i.e., number of radioactive decays per second) radium is about 100 times worse than plutonium if ingested into the body.[10] This follows primarily from the fact that almost all ingested plutonium is excreted, as previously mentioned in Chapter 11, but radium is not. Thus, if one adds 0.1 percent to the earth's radioactivity, one adds much less than this in terms of danger.

NUCLEAR WASTES WILL NOT CHANGE
THE EARTH VERY MUCH

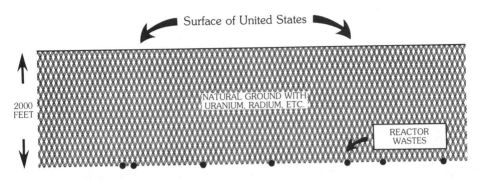

There are 1,000 times more cancer doses from natural radioactivity in the ground (down to 2,000 feet) than in the annual waste burial from 400 reactors.

Other Comparisons:

- The reactor wastes will be solid and insoluble
- The reactor wastes will be in dry places.
- The reactor wastes will become another 1,000 times less lethal after 400 years.

Perspectives on waste disposal.

Other comparisons can be made, for example between the wastes put out by a given reactor each year and the amount of uranium ore mined to fuel that reactor. After 300 years, the "radiotoxic hazard measure" is less for the wastes than for the ore.[5,6] This means that if the wastes and ore were dissolved in water, and the same fraction of each were ingested, fewer cancers would be produced by the waste than by the original ore which produced it.

This might sound as if we are cleaning up the earth by taking the ore out of the ground and burning the uranium in reactors. This is not the case, at least not for thousands of years. Although the wastes are less hazardous after 300 years than the original ore, the chief hazard of the ore—in the form of radium—is still with us because it was removed from the uranium before the fuel elements were fabricated.

It is clear from this discussion that we are doing nothing very drastic toward long-term contamination of the earth by depositing nuclear wastes deep underground. However, for a substantial period of time, certainly for several hundred years, the wastes would be hazardous unless handled with care. For this reason, the disposal sites should be well chosen. Salt mines are often considered for this purpose. Partly as a test of the handling equipment the AEC proposed the use of salt mines near Lyons, Kansas, in the belief they would be free of water. Later, it was determined that old mine borings negated the integrity of the Kansas site. Nevertheless, given an undisturbed site, salt deposits have many advantages for waste disposal.[11]

It is quite probable that the Lyons experience accounts for the present caution exercised in reaching final decisions. Perhaps the suggestions to store the wastes in retrievable canisters are a mechanism for putting off the day of decision a few decades, particularly since it is possible constructive uses may be discovered for the wastes in the future.

Salt deposits underlie hundreds of thousands of square miles of the U.S. Because of that fact, it should be a simple matter to find a satisfactory site a few square miles in area—or, in fact, find many diverse sites. Moreover, salt deposits are geologically among the most stable and water-free formations known.

Underground salt deposits would be unsatisfactory only if geological or other processes exist that could cause the stored material to be brought up to the earth's surface, some 2,000 feet, and then to be dispersed so that somehow it finds its way into human bodies. B. L. Cohen[6] has studied the problem in detail. Fortunately, geologists and other scientists know a great deal about the movement of ground water, the leaching rates of glass-like material, ion-exchange processes, and others. The natural barriers to such dispersal are extremely large. About the only way that wastes buried in salt could present a problem would be if they were dug up during salt mining. Cohen here concludes that even if society forgot where the wastes were buried, there is so much salt in the United States that the chances of accidentally digging up the wastes are vanishingly small.

If salt-mine or other geological disposal were still to be regarded as undesirable for some reason, a point made by Chauncey Starr and R. Phillip Hammond[12] appears convincing—namely, that if all else fails, wastes could still be stored quite safely in structures like the Egyptian pyramids. They suggest a structure on the scale of the great pyramid of Cheops could accommodate ". . . all the nuclear wastes that could be generated by the United States at its present rate of electric power consumption, for over 5,000 years."

Other suggestions have been made which are also probably feasible.[7] It is unnecessary to detail them here.

It is evident that the long-term storage problem is a problem only because several satisfactory choices are available but no selection has been made. The very existence of the choices brings forth both indecision and criticism!

Accounts of the leaks of radioactive materials at the Hanford nuclear-waste facilities leave the impression that a startling degree of carelessness was involved. Liquid wastes, generated in the nuclear-weapons program during and soon after the Second World War, not only leaked from the containment tanks, but some of the leaks were not discovered for several months. It is not surprising that leaks occurred; the most spectacularly offending tank was built in the haste of the war. But it is hard to excuse the failure to replace the tanks at an earlier time or to institute procedures which would have detected leaks more promptly. Nevertheless, the leaks resulted in injury or illness to no one, and no indication has been found that anyone will be harmed in the future.[13]

Perhaps an analogy is in order to provide a reasoned response to the Hanford incident. If, in the early days of civil aviation, a propeller fell off a plane but the pilot landed safely and no one on the ground was struck, one would have been distressed at the bungling involved and would have been determined to avoid similar mishaps in the future. But it would have been absurd to shut down civil aviation.

Similarly, accidents stemming from mishandling of the weapons program wastes can serve as reminders of the need for meticulous care in handling the wastes in the more deliberately developed civilian-power program. But it is difficult to see why, as critics have charged, the instances of past carelessness should be viewed as positive indication that comparable mistakes will be made in the future. Civil aviation has been made safe—at least safe enough for public acceptance—and similar care can be demanded and imposed elsewhere. Indeed, the liquid wastes at Hanford are now being solidified in an apparently satisfactory manner.[14]

Finally, we face the issue of concern for future generations and the possibilities of sabotage. Some have argued that, even if the time span of danger is reduced from 100,000 years to 600, the demands on human vigilance and social stability would be unreasonable. But even if wastes were to be stored above or below ground in accessible man-made structures, all that is really being asked of future generations is that they do not attempt to destroy the canisters with bombs (in a high level of civilization) or, if it were possible, with pick axes (in a low level of civilization).

We make much greater demands on the good judgment and stability of future generations when we provide society with nuclear weapons, chemical explosives, cities with no internal food supply, and water supplies that need constant attention. As for sabotage of storage facilities, the inherent difficulties facing the would-be saboteurs are much greater than are presented by many other much more vulnerable features of our society.

The issue of responsibility to future generations cannot be simplified to the point at which the only danger considered is the extremely remote one derived from radioactive wastes. Our present use of fossil fuels represents a dissipation in a few hundred years of an accumulation made by Nature over millions of years! It seems that is an instance of far greater disregard of the needs of future generations than the storage of radioactive waste in a few selected—and safe—sites.

REFERENCES

1. Allen V. Kneese, *The Faustian Bargain,* Resources, p. 2 (Sept. 1973). (Resources for the Future, Inc.)

2. See T. H. Pigford, Annual Review Nucl. Sci. *24,* 515 (1974) for an excellent summary of steps in the nuclear fuel cycle.

3. See William Brobst, Nuclear Technology *24*, 343 (1974) for a typical description.

4. A. S. Kubo and D. J. Rose, Science *182*, 1205 (Dec. 21, 1973).

5. J. Hamstra, Nuclear Safety *16*, 180 (March-April, 1975).

6. B. L. Cohen, Physics Today *29*, 9 (Jan. 1976) and *High Level Radioactive Wastes from Light Water Reactors* (1975), to be published.

7. *High Level Waste Management Alternatives*, WASH-1297 (May 1974).

8. Frank N. Pittman, Nuclear Technology *24*, 273 (1974).

9. Activity in wastes based on *High Level Waste Management Alternatives*, BNWL-1900 (May 1974) and on Ref. 2; uranium in earth based on Ref. 6.

10. See Ref. 5 and sources cited therein.

11. *Disposal of Solid Radioactive Wastes in Bedded Salt Deposits*, NAS-NRC Report (November, 1970).

12. C. Starr and R. P. Hammond, Science *177*, 744 (1972).

13. Robert Gillette, Science *181*, 728 (Aug. 24, 1973).

14. Seattle-Post Intelligencer, Oct. 4, 1974.

ARE NUCLEAR REACTORS SAFE?
Chapter Thirteen

Both nuclear reactors and nuclear bombs produce energy by a chain reaction propagated by neutrons. Consequently, it is not surprising that some persons believe a nuclear-power plant could go out of control and explode like a gigantic atomic bomb.

Such a catastrophic event in today's commercial reactors, however, is not possible for a number of reasons, chief among which are these:

- The U-235 or Pu-239 in a nuclear reactor is not sufficiently pure to produce an explosion.
- The average time in a reactor from the birth of a neutron to the point at which it is reabsorbed to create a new fission event is much too long to cause an explosive reaction.

Those are facts of physics and of nature. The fact is no one, not even the most outspoken foes of nuclear power, has seriously argued that a reactor is a giant atomic bomb.

On the other hand, if the radioactive products in the fuel core were somehow dispersed in the atmosphere by, for example, a malfunction of the reactor, the radioactivity would indeed be lethal to anyone receiving sufficient radiation exposure. It is the possibility of this event that has drawn considerable attention and promoted an understandable fear and opposition to the deployment of nuclear power.

In order to consider evidence that the danger has been greatly exaggerated, it is necessary to outline some new technical considerations.

The principal reason a means exists for spreading the radioactivity of the core is that the radioactivity itself produces heat even after the reactor has been shut down. Initially, the rate of heat generation is as much as 7 percent of the total reactor power; for a power plant producing 1,000 megawatts of electric power, then, the evolution of residual heat amounts to about 200 megawatts (when the efficiency of conversion from thermal-to-electrical energy is taken into account).

As isotopes with short half-lives decay away, the rate of heat generation drops very rapidly at first, so that it is about half as much in only 45 seconds. However, the radioisotopes with longer half-lives then begin to dominate, and the heat rate remains great many hours later. Under steady operation, most of the decay heat is utilized in power generation.

If the normal flow of cooling water continues, a shutdown of the reactor creates no difficulty. Unexpected shutdowns can occur quite frequently even in regular operation—as, for example, in the case of lightning causing an electrical circuit-breaker to trip. However, if the water were lost by, for example, a ruptured cooling pipe, the residual heat in the fuel elements could be sufficient to melt the core structure and, thus, possibly cause release of the radioactive products to the environment.

Loss of water itself stops the chain reaction because the water is needed as a moderator of the neutrons. It is only in the event of a core melt that a danger might arise of a significant release of radioactivity that could cause injuries or deaths outside the power plant. But even that occurrence is highly unlikely, because the designers of nuclear plants foresaw the remote danger and placed the entire reactor inside a steel-reinforced, thick-walled concrete building.

If the normal cooling process were to fail, back-up safety cooling systems are provided to take over and assume the task of keeping the hot core from reaching the melting point.

In the worst situation envisaged, one of the main large cooling lines—a stainless steel pipe about three feet in diameter with three or four inch thick walls—conceivably could break in two places simultaneously. All the steam and water in the reactor vessel would then burst out into the reactor containment building, leaving the core vulnerable to a "meltdown."

A final emergency core cooling system (the ECCS) is designed then to flood the reactor vessel with water.

But let's go even further. Suppose the ECCS should also fail? Other emergency systems would come into play, including a massive building sprinkler system, to minimize the effects of a possible core melt.

Someone is bound to suggest those systems could fail, too. Is it conceivable that all safety systems could fail simultaneously? Anything is *conceivable*, but is the chance large enough to worry about?

To answer the question is not easy. Since no such accidents have occurred —indeed there haven't been any core melts in reactors of the light water type—the question can only be answered by *calculating* the probability of occurrence. For the moment we will defer discussion of how this can be done.

It is easier to examine first the actual operating records. There are over 55 such reactors in commercial use in the United States today. Many have been in operation for five or more years. There are over 70 commercial reactors outside the United States, many of which are similar in design to our LWR's. Others are of diverse types. There have been no reports of core melts or radiation-related injuries in any of these reactors. These data include about five commercial reactors in the USSR.[1]

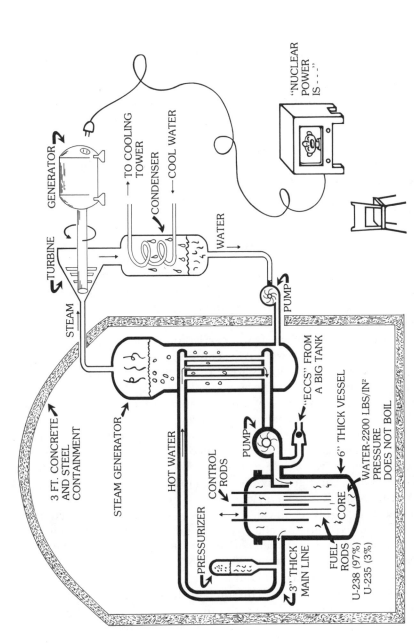

A simplified schematic diagram of a common type of light water reactor (LWR): the Pressurized Water Reactor (PWR). The principle of operation is very simple: When the control rods are pulled up, the chain reaction starts and heats the water in the reactor vessel. A real reactor of commercial size has four "main lines" for cooling water and four steam generators. The Emergency Core Cooling System consists of several sub-systems. Shown schematically—with a "passive" check valve—is the one which is discussed in Appendix D.

GENERATOR

TURBINE

→ TO COOLING TOWER

CONDENSER

← COOL WATER

"NUCLEAR POWER IS ---"

STEAM

WATER →

3 FT. CONCRETE AND STEEL CONTAINMENT

STEAM GENERATOR

PUMP

HOT WATER

"ECCS" FROM A BIG TANK

CONTROL RODS

PUMP

6" THICK VESSEL

PRESSURIZER

WATER-2200 LBS/IN² PRESSURE DOES NOT BOIL

CORE

FUEL RODS
U-238 (97%)
U-235 (3%)

3" THICK MAIN LINE

Outside the commercial reactors, the record, while very good, is not un-blemished. Fifteen years ago three technicians were killed in an accident with an army test reactor. While performing maintenance work, one of them inad-vertently *withdrew* a control rod manually, causing a sudden rise in reactivity. It was like turning left into oncoming traffic instead of turning right.

On the other hand, there are 112 U.S. naval ships—mostly submarines—powered by nuclear reactors similar to the commercial pressurized water type. In total, these ships have operated (through mid-1975) for a collective total of 1,300 years—all without a core melt or a radiation-related injury.[2] It can be said—if all factors were taken together—that the total operating experience without a core melt in light water reactors is equivalent to about 2,000 years of a single reactor; the experience to date, then, is 2,000 reactor years.[3] Thus, the operating experience itself tells us that the chance of such a failure for one reactor is less than about once in a thousand years. Another way of expressing the same thing is with the statement that the chance of failure is *less* than one in a thousand per reactor per year.

That limit is based upon an empirically determined fact. However, since *no* failures have occurred, it cannot be surmised from that limit whether the actual chance of failure per reactor per year is one in a thousand or one in a billion!

To obtain a better estimate of the safety of reactors, a detailed study of the design features is required. Accordingly, the AEC initiated such a study in 1973 under the direction of Professor Norman Rasmussen of the Mas-sachusetts Institute of Technology. It was carried out by a group of 60 scien-tists and engineers, 10 of whom were AEC employees who assisted the "inde-pendent" group. Various consultants were called in. The initial study cost about $3 million. The results first became available in draft form in 1974,[4] and a final publication was purposely put off until the general public, friend and foe alike, had a chance to analyze or criticize points in the report.

After about 16 months of critical review, checking of calculations, and incorporation of other various refinements (all of which cost another million dollars), the final report was made available in October 1975. It is entitled "Reactor Safety Study," and goes by the designation WASH-1400.[5] It is also often referred to as the Rasmussen Report.

The report itself is highly technical. It consists of a summary, followed by a main report, which explains the chief results and how the calculations are made, and then 11 appendices which contain the intimate details of the data base and the calculations. In all some 3,300 pages!

It is the conclusions which are, of course, our principal interest, so we will discuss them first. Later, we will explain some of the methods used, the criti-cisms leveled by critics, and conclude with our own assessment.

First of all, the report considers a wide spectrum of possible "scenarios" which could produce a core melt. In place of the empirical (or actuarial) number mentioned above—that is, less than one chance in a thousand per reactor per year for a core melt—the study predicts one chance in 20,000 per reactor per year.[6] Moreover, the possible *consequences* of a core melt are

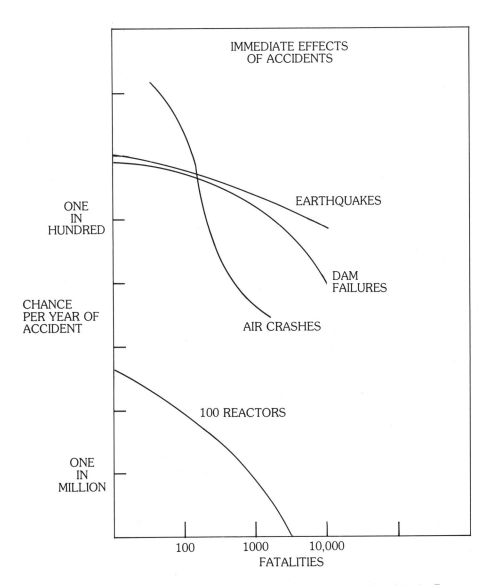

Comparison of accident chances for reactors and other causes, as calculated by the Reactor Safety Study. The curves indicate the chances of causing the indicated number of fatalities, or more, in any given year with 100 reactors operating. For example, the odds are about one-million-to-one against 1,000 or more people being killed in a reactor accident, but not even one-hundred-to-one against a dam failure with the same number of fatalities. (Adapted from WASH-1400.) NOTE: Moving one division along vertical scale represents changing chance by 10.

analyzed in detail. Contrary to much previous opinion, the chance of a core melt causing deaths or injuries or genetic damage to the public is found to be extremely small. Indeed, there is only one chance in three million per year that a nuclear plant would cause ten or more fatalities.[7]

There are other ways to show the smallness of the risk. At the end of 1975, about 55 reactors were in operation in the U.S., but in four or five years the number is expected to rise to about 100. The Rasmussen study indicates the chance an average U.S. citizen has of meeting death through an accident to a nuclear reactor when 100 of them are operating is one in 50 billion per year.[8]

By comparison, a person who never rides in an automobile and never crosses a street nevertheless has one chance in 300,000 each year of being killed by a runaway automobile![9] In other words, the average person has 100,000 times as much to worry about from runaway cars as from 100 nuclear reactors. By any reasonable standard one can consider the risk involved from a nuclear plant as being absolutely negligible.

But what are the odds if one lives very close to a nuclear plant? Suppose you lived right at the boundary, and you did not "evacuate" if there *were* an accident. Even then, the predicted chance of being killed is only about one in 10 million per year[10]—which is still about 300 times smaller than for that devotee of super-safety who avoids all deliberate contact with automobiles. In fact "Super Safety" has about the same chance of being killed by a plane that falls on him because 20 such deaths occur each year in the U.S. on the average.[11] Even hydroelectric power is more dangerous than nuclear power; the average American's chance of being killed by a dam failure is over 1,000 times greater.[12]

The Reactor Safety Study also shows that risks of death in many other man-made activities are much higher than from 100 nuclear reactors. They include falls, fires, drowning, falling objects, and electrocution.

When comparisons are made with natural phenomena—lightning, tornadoes, hurricanes, and falling meteors—for these only the risk of being killed by a meteor is less than that for 100 nuclear plants. In fact, the chance that a giant meteor will get through the atmosphere and hit a city, killing 1,000 people, is about the same as the chance that 100 nuclear plants would produce the same number of deaths—one in a million per year.[13]

Another finding of the Rasmussen investigators is that the chance of a nuclear accident being caused by an earthquake, a flood, a storm, or a plane crash is even smaller than by internal plant failures.[14]

What about effects due to radiation released to the environment which does not produce prompt death, but nevertheless might create a long-term potential for fatal cancer or genetic damage? This question is the most complex, so it will be given more attention later. Suffice it to say at this point, that there is about one chance in a million per reactor per year of an accident which might produce as many as 200 latent cancers per year;[15] these would continue over a period of about 30 years.

If nuclear power is permitted to become a mainstay of American energy production in the final quarter of the 20th Century, it is expected 1,000

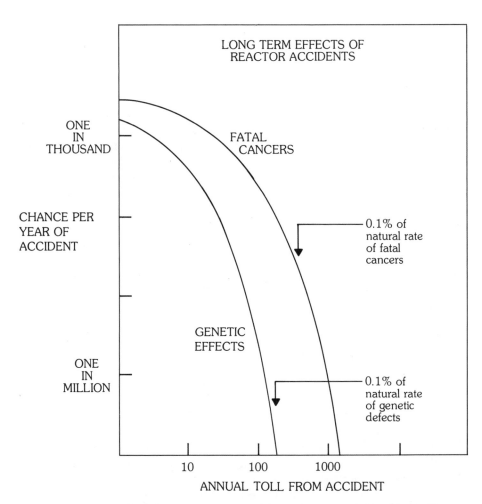

Cancer production and genetic damage from nuclear accidents, as calculated by the Reactor Safety Study. The curves indicate the annual chance for causing the indicated effects, or worse, with 100 reactors operating. For example, the odds are about one-million-to-one against producing 1,000 cancers per year (lasting for 30 years) or 100 genetic effects per year (lasting for several generations). These rates are about 0.3 percent of the present national cancer rate and about 0.1 percent of the present rate of genetic defects. (Adapted from WASH-1400.) NOTE: Moving one division along vertical scale represents changing chance by 10.

reactors will be in operation by the year 2000. What will be the risk then? The Reactor Study points out even greater safety is already being designed into the new and projected reactors, so that it is not valid to extrapolate to 1,000 reactors,[16] because the risk per reactor will probably decrease. Other technology shows a similar improvement; for example, the risk of death per mile associated with automobile travel has decreased about threefold in the past 40 years.[17] Nevertheless, the 1,000 reactors would cause far less danger than any of the other man-made sources of risk discussed—even without the expected improvements.

Two questions remain: First, if nuclear power is so safe, why did the issue arise in the first place? Second, should we believe the results of the Reactor Study? The first is easy to answer, the second more difficult.

In 1957, before any commercial nuclear power plants were in existence, the AEC made a study (WASH 740) to determine the magnitude and consequences of a *worst case accident*. It is somewhat analogous to investigating the worst possible consequences of, say, the Empire State Building falling over—without considering the likelihood of such a thing happening; the 1957 study did not address itself to the question of what the chances actually were for such an accident. A "worst case accident" means one in which all reactor safety systems fail simultaneously, along with the worst possible weather conditions and the densest population, which was assumed to be immobile.

The 1957 study had little data to go on, since the safety systems were scarcely designed, let alone built and tested. Moreover, no attempt was made to assess the probability that such combinations of circumstances could occur. Thus, it is not surprising that the 1957 study predicted that a catastrophic accident in even a small reactor (150 megawatts of electricity) could cause 3,400 fatalities and 43,000 serious illnesses.

As a result, it is easy to understand why the 1957 study generated a genuine concern over the safety of reactors, particularly when the numbers are multiplied by about 7 to have the figures match the much larger reactors of today, an extrapolation which the later Safety Study clearly shows to be not valid.

That should answer the first question. We turn to the second: how reliable is the Rasmussen Report? We will attempt to answer by discussing first how the Safety Study made its predictions.

First, we note that in the years between 1957 and 1972, reactor engineers designed new safety systems, but some concerned citizens questioned the reliability of these systems, chiefly the emergency core-cooling system, or ECCS,[18] thus setting the stage for the controversy over safety.

Reliability is usually determined by direct test. So, for example, engineers in the automobile industry test thousands of valve springs for engines to determine the reliability of the springs. A test on one spring can provide a "yes" or "no" answer on that spring alone; while the test is not meaningless, it tells us very little, statistically speaking. Similarly, a complete test of the ECCS would require severing one of about eight steel cooling pipes with a single, guillotine-like blow and removing a section; each of the pipes is some 3 feet in diameter and 3 inches thick, and the eight pipes comprise the main cooling system of a big reactor. Again, one such test would tell very little about the chance of failure of the ECCS in many reactors unless the test uncovered a fundamental design flaw. Even the non-specialist can appreciate the magnitude of such a massive and destructive experiment. Appendix D discusses the problem in more detail.

A far more practical method is to study the past record of performance of similar steel pipes in similar service, like those in oil pipelines or coal-fired plants, and to combine those performance records with information for the components of the ECCS and other safety systems. That would determine a combined chance of failure. It is a perfectly logical approach. The principle can be illustrated by the following data drawn from personal experience:

One of the authors and his wife have driven automobiles approximately 700,000 miles. The time required was about 20,000 hours, or about 10 million time intervals, each one 7 seconds long. The reason for choosing 7 seconds as a time interval will be clear as the explanation continues. In all the driving, four instances were recorded in which rear tires suffered total and almost instantaneous failures; that is, blowouts. Thus, by experiment the chances are about 4 in 10 million of total failure in rear tires during any single 7-second interval.

In those same 700,000 miles, one sudden and complete failure of a steering gear occurred, which meant the chance of such a mechanical failure was about one in 10 million per 7-second time interval. Fortunately, neither the blowouts nor the steering-gear failure produced any serious consequences.

Let's suppose, however, that they had occurred virtually simultaneously, or within the same 7-second interval. If the time interval were greater than 7 seconds, it can be presumed the driver might take adequate independent corrective measures for each event to avert tragedy. On the other hand, it can be assumed that the consequences of a simultaneous occurrence could indeed be serious.

To determine the chance of such a coincidence by experiment is essentially impossible, but it can be calculated. The way to do it is to multiply one probability by the other—for example, the probability of steering-gear failure times the probability of tire failure.

Thus, the combined probability can be figured this way: One divided by 10 million times four divided by 10 million—or four in 100 million million per 7 seconds. Multiplying by the 10 million time intervals gives four chances in 10 million of simultaneous failure during our total of 700,000 miles of driving. Put another way, it says we would have to drive a distance of about 700,000 times 2.5 million miles to make a *probable* observation of one such serious double event.

Obviously, it's a hopeless task. Moreover, it can be concluded that we need not fear realistically that such a simultaneous event will take place.

To improve the accuracy of our calculation, we could collect information from many drivers so the input data would become more precise, but it still might be impossible to observe a significant number of double events. Nevertheless, our ability to predict the probability would become more accurate.

The illustration can be carried further to illustrate another principle. It might be agreed that the front tires were as likely to rupture suddenly as the rear tires. and that it was only by chance that it was the rear ones that failed. Now we are faced with a new possibility: Whatever caused a front tire to fail might also cause a steering-gear failure. An obstruction in the road could have been the cause.

That type of a double failure is referred to as a common mode failure. Still another possibility is that a front-tire failure itself could bring on a steering-gear failure, or vice versa, thus producing an event link, or an "event tree." The analysis, then, becomes more complex because one must consider the probability of the tire failure triggering the steering-gear failure, and this can be done by detailed analysis of the design of the steering gear or by actual tests.

Once an event has been created either singly or in a common mode group of failures or triggered by events linked in a tree, it must be inquired next what the possible consequences are. For the case at hand, the driver may take corrective steering action or protect himself in advance by fastening his seat belt, etc. These measures may also fail, and, of course, they have a specific chance of failing. For example, the driver may forget to fasten his seat belt once in ten times. That introduces the "human fault tree."

Finally, the situation that may exist outside the car must be considered. Is an obstruction lying in the path of the car? Does the weather create additional hazards? From these and other data an observer can calculate the chance of one or another kind of accident or other consequences—as, for example, the chance of the driver being killed or injured or the possibility of producing external property damage.

The study described in the Rasmussen Report employed just such methods, but the problems are more complex, requiring an intimate knowledge of the details of reactor design, the failure rates of each component in the reactor and safety systems, the weather patterns, the population distribution near the reactors, and many more factors.

An analogy with the double failure of the tires and steering gear would be a double failure of both the cooling pipe *and* the ECCS. The Rasmussen Reactor Safety Study identified and screened thousands of such potential accident paths!

Thus, each reactor accident path leads to a predicted probability for one or another of various possible consequences. These range from trivial to serious, just as our tire-steering-gear failure could be trivial or serious. It is these final predicted consequences, expressed in the number of people killed or injured, the extent of property damage, etc., that produce the final relevant results of the report. We have already quoted some of them, together with comparisons with other risks of living.

It is extremely important to understand that the natural and other man-made risks are based almost entirely on real or *actuarial* data. Since no commercial reactor accidents leading to death or radiation injury have ever been recorded, the reactor results from an accident are arrived at by calculation.

Just as the automobile example has a large uncertainty (only *one* steering-gear failure), so, too, do some of the predictions of the Reactor Study (large steel pipes do *not* fail very often, so the data "base" is not very great).

However, just as the automobile example leads to so small a result that it is not important whether the error is tenfold or more, so, too, does the Reactor Study lead to some results with large but unimportant uncertainties.

In its final pages, the Reactor Safety Study states:

> "We do not now, and never have, lived in a risk-free world. Nuclear accident risks are relatively low compared to other man-made and natural risks. All other accidents, including fires, explosions, toxic chemical releases, dam failures, earthquakes, hurricanes, and tornadoes, that have been examined in this study are much more likely to occur and can have consequences comparable to or greater than nuclear accidents."[19]

Should we believe these words and the results quoted earlier? We, the authors of this book, do and we will amplify our reasons below. There are others who do not.[20] Shortly after its issuance, Ralph Nader labeled the report "fiction."[21] As scientists who have devoted many years to research and development, we cannot accept this notion that 60 reputable fellow scientists and engineers were so dishonest as to produce a fictional report on so vital and sensitive an issue. We can only presume that Nader has consulted people *he* considers to be expert and join in his judgment.

It is difficult for any individual, whether he be layman, physicist, or engineer outside the specific field of nuclear-reactor safety, to evaluate fully every detail of the Reactor Study. Ultimately, in this and other areas—for example, the safety of a new airplane—one must make a judgment: Which of the "experts" appear to have studied matters most carefully and considered them most responsibly?

In our reading and analysis of the Rasmussen Report, we found much to give us assurance that in their two-year study the 60 scientists and engineers clearly brought their most sincere and honest expertise to bear on this monumentally-important problem of nuclear-reactor safety. The fact that they were paid by the AEC, and later by the Nuclear Regulatory Commission, to do the study is to the credit of the AEC and does not, in our view, mean that their findings are to be discounted. In so significant a national issue, wasn't the mounting of such a study substantially what science itself and the American people would have demanded of a responsible federal agency? In fact, it provides an example that other agencies at all levels of government might duplicate to everyone's advantage.

The Rasmussen Report is not unique, among expert studies, in its encouraging evaluation. What is perhaps an even more optimistic evaluation came out of a study commissioned by the Swedish government.[22] Issued in June, 1974, it was undertaken to determine the advisability of placing nuclear-power plants close to cities so that the waste heat could be utilized for urban-heating purposes.

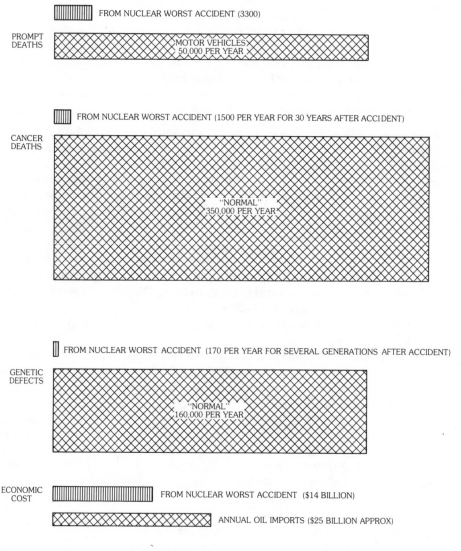

THE NUCLEAR WORST CASE ACCIDENT IS EXPECTED ONCE IN 10,000,000 YEARS
(WITH 100 REACTORS)
THE "NORMAL" TOLL COMES EVERY YEAR

FROM NUCLEAR WORST ACCIDENT (3300)

PROMPT
DEATHS

MOTOR VEHICLES
50,000 PER YEAR

FROM NUCLEAR WORST ACCIDENT (1500 PER YEAR FOR 30 YEARS AFTER ACCIDENT)

CANCER
DEATHS

"NORMAL"
350,000 PER YEAR

FROM NUCLEAR WORST ACCIDENT (170 PER YEAR FOR SEVERAL GENERATIONS AFTER ACCIDENT)

GENETIC
DEFECTS

"NORMAL"
160,000 PER YEAR

ECONOMIC
COST

FROM NUCLEAR WORST ACCIDENT ($14 BILLION)

ANNUAL OIL IMPORTS ($25 BILLION APPROX)

The "worst case" nuclear accident compared to "normal" life. (Data is from WASH-1400, ref. 5.)

Sweden's study committee paid much attention to the question of major accidents. It concluded that in the worst conceivable case, several hundred fatalities might result, but that was for a reactor located 5 kilometers (3.1 miles) from the center of a city with a population of a million people, with a restricted population zone around the reactor of only 0.5 kilometers. The chance for such an accident was estimated to be one in 10 billion per reactor year.[22] For population restrictions extending to 2 kilometers from the reactor, the number of fatalities in the worst conceivable accident is reduced to "a few cases." The committee found no reason not to adopt the larger safety zone of 2 kilometers, nor would there be any reason not to do so in the U.S. In fact, present-day practice in the U.S. is far more cautious than that.

Several qualifications should be noted in citing the Swedish Report. Although the plants considered were of the light-water type that dominates American planning today, their power output was about half as great as those planned in the U.S. Therefore, in a given accident example, the radiation dosage rates should be assumed to be double, and the regions of population exclusion should be increased somewhat to maintain the same level of safety.

In addition, the Swedish study represents the views of the Swedish energy establishment, much as the American report might be construed as representing the views of the American energy establishment. Just as the Rasmussen Report has been received with criticism here, so has the Swedish Report been the subject of criticism in Sweden. Nevertheless, Americans may take some comfort from the fact that our experts are not alone in their assessments.

The above discussion of possible accident sequences in a reactor accident has concentrated on the loss of cooling water, followed by the failure of the emergency core cooling and other safety systems, with subsequent escape of radioactivity from the building containing the reactor. Another conceivable accident, which has been the subject of study for many years, should be considered—namely, the rupture of the reactor pressure vessel itself.

Although it may sound like an ominous possibility, the various reactor-safety studies (including the Rasmussen Report and the American Physical Society Study to be taken up here) agree it is not a serious danger. Substantial experience with high-pressure vessels fortifies and justifies that conclusion.

The pressure vessel of a large reactor is truly a formidable steel structure. Its walls are 6 to 10 inches thick, and it is made of steel specially chosen for its strength and ductility at high temperatures. Its internal integrity is studied in the same way a radiologist searches for suspected cracks in human bones—with X-rays. In operation at a nuclear plant it is subjected to stresses no greater than those experienced by the steel used in boilers of modern coal-fired power plants, in great suspension bridges, and in modern skyscrapers. We trust our lives without hesitation to the durability of these structures, and no basis exists for doubting the integrity of the much more carefully planned and examined nuclear-reactor boiler tanks.

Since the Rasmussen Reactor Safety Study first appeared in draft form for comment, many individuals and groups have publicly assessed its results, some reaching conclusions that are not in agreement either with the Report or with each other—and which are themselves open to question. The final report took cognizance of the criticisms, and, in some instances, the criticisms led to constructive changes.[23]

One of the strongest negative criticisms came from the Sierra Club, in cooperation with the Union of Concerned Scientists.[24] They produced a study whose chief conclusion is that the Rasmussen Report underestimated the probabilities of reactor failures by a factor ranging from 10 to 100, depending upon the severity of the accident. *Even with this large increase* (which has provided the rationale for attacks on nuclear power, such as those made by Nader) reactors are very safe. For example, the chance for a catastrophic accident remains smaller than for hydroelectric-power dams, whose safety is accepted with virtually no concern or dissent! The main basis for the Sierra Club's bigger probabilities was its contention that the method of calculation was not valid.

Professor Rasmussen answered the critics of the "methodology" in pointing out that in England these calculational methods have been in successful use for about ten years for predicting other kinds of accidents. Interestingly, the predicted accident rates turned out to be a little greater than the actual accident rates.[25]

The worldwide interest in reactor safety is illustrated by a lengthy critical analysis of the Sierra Club report made by Norwegian experts. They conclude that the Sierra Club report contains a number of errors which cast serious doubt on the report.[26]

The U.S. Environmental Protection Agency made an assessment of the Rasmussen study and generally endorsed its methodology and conclusions, with the exception that the EPA believed the possible consequences of a major release of radioactivity had been underestimated.[27]

Meanwhile, the American Physical Society, the leading professional organization of the physics community, appointed a group to make an independent reactor-safety study. The chief findings of the APS Study merit a review in some detail.[28]

As an over-all conclusion, the APS Study states it has "not uncovered reasons for substantial short-range concern regarding risks of accidents in light-water reactors." The implication is that a basis exists for long-range concern. In that context, the APS Study emphasizes the need for continued research and development in safety systems and, above all, the need for maintaining high standards in manufacturing and personnel training. Satisfying the latter need does not appear to be an insuperable problem.

The situation is similar to that of the aircraft industry, which has also been required to maintain a very high quality of manufacturing, inspection, and personnel training. Air travel has become much safer per passenger mile over the years—rather than more dangerous, a condition that might exist today if the industry had become complacent as growing air travel became routine.

In fact, the nuclear-power industry has started with a much better safety record than the aircraft industry has had. Despite the good record to date, the safety of reactors will probably become greater through constantly upgraded safety measures stimulated by the inevitable small failures that have occurred and will continue to occur. These small failures have brought no injuries or deaths and perhaps never will, but they can provide continuing reminders of weaknesses in the systems.

The APS Study did not make an independent assessment of the probabilities for various kinds of possible reactor failures. That is unfortunate, perhaps, since the method of "event tree" and "fault tree" analysis utilized by the Rasmussen Report has been the subject of criticism by some other groups. It is especially unfortunate because the probability analyses were the major advance made by the Rasmussen Report over the 1957 study.

On the other hand, the APS Study does indicate specific areas in which improvements in these analyses can be made. In consequence, the present reactor-safety research program undertaken by the federal government, for which over $50 million was budgeted in the year 1976, can take guidance from the APS suggestions.

A substantial portion of the APS Study is devoted to a recalculation of the casualties which might result from a major release of radioactivity in a nuclear-plant accident. While agreeing with the Reactor Safety Study calculations as far as they were carried in the draft issue, the APS Study extended the area subjected to hypothetical long-term radiation effects. These are due to very low levels of activity from long-lived radioisotopes that might be carried long distances by winds.

The number of people exposed in the hypothetical accident considered by the APS Study is about 10 million. With such a wide dispersal of radioactive debris, extending out for 500 miles in the direction of the wind, the radiation level produced beyond a few miles is very low. It is estimated to increase the chance of a person's dying of cancer during his lifetime by one-tenth of a percent, thus resulting in 10,000 possible additional cancer deaths in the 10 million exposed people. We say "possible" because it has not been established that cancer is produced at these low radiation levels.

The estimated cancer deaths were calculated by assuming that the "linearity hypothesis" is valid. Under this hypothesis, or assumption, knowledge gained in studies of the effects of intense radiation exposures over short periods of time is extrapolated, or projected mathematically, to infer the presumed small effects produced by low levels of radiation extending over any period of time. In the absence of better experimental data it is a prudent practice, particularly for establishing radiation limits for the general public, but most experts agree it over-estimates the possible effects. In view of the importance of the linearity hypothesis it is discussed in detail in Appendix C. Other assumptions which tended to maximize the possible casualties were made in the APS Study, as for example, a steady wind with no rain.

While the prospect of 10,000 additional cancer deaths should not be viewed with complacency, even though spread over several decades, it is important both to place the absolute number in proper perspective and to bear in mind that the chance of the initiating accident is very remote. As recalculated in the revised Rasmussen report, that chance is about once in 3 million years per reactor,[29] or less than the probability above of a double failure of a tire and steering gear.

Further perspective is gained in noting that the average person's chance of dying of cancer from other causes is nearly 20 percent; even the change in the cancer rate in a year's time is one or more percent.[30] Thus, a possible change by one-tenth of one percent because of a nuclear-reactor accident would be very difficult and perhaps impossible to detect.

The final report of the Reactor Safety Study—The Rasmussen Report— took cognizance of the American Physical Society extension. Additional expert advice was called upon, particularly in assessment of the latest data concerning the biological effects of radiation, resulting, in effect, in a slight softening of the linearity hypothesis.[31]

The final prediction for long-term cancer possibilities due to widespread distribution of low levels of radioactivity were altered from the original draft. In the worst conceivable accident, having a probability of one in 10 million per year for 100 reactors, the final draft predicts that widely spread radioactivity could produce 1,500 latent cancers per year over a 30 year period.[31] These delayed cancer deaths would be in addition to about 3,000 prompt deaths produced in the worst case accident. For comparison, the normal incidence in the affected population is 17,000 per year. Thus, bad as this unlikely event would be, it is hardly the holocaust some people imagine.

The Safety Study considers many other possible aspects of reactor accidents, as for example, property damage, area requiring decontamination, reduction of risk by evacuation, time available for evacuation, etc. We shall not attempt to summarize these in any detail. Most noteworthy, however, is that in every case, the probabilities for harmful effects are very low.

After the publication of the preliminary draft of the Safety Study, an accident occurred at Browns Ferry on March 22, 1975 involving two large reactors operated by the Tennessee Valley Authority (TVA).[32] While searching for air leaks in the containment building using a candle flame. a workman accidentally set fire to material surrounding electrical wiring. Ironically, the accident was the result of a safety feature of reactors: the containment buildings are maintained at a slight constant negative air pressure to prevent possible leakage of radioactivity.

The ensuing fire did considerable damage to electrical wiring. In particular it made some safety features inoperative. No radioactivity escaped, and no one suffered radiation injury. The largest cost incurred by TVA is the cost of substituting alternative electrical energy during the repair of the reactors.

The accident was fully investigated by the Nuclear Regulatory Commission. Its findings are available in an 1,100 page report.[33]

The Browns Ferry accident has been termed "the accident which couldn't happen." However, the Safety Study, post-facto, analyzed the accident. It concludes that there was only one chance in about 300 that remaining back-up systems might have failed, thus leading to a core melt.[34] When we recall that the chance that a core melt leads to serious consequences is less than 1 in 10, the chance that the Browns Ferry accident might have resulted in even a single death is less than one in 3,000.

It is sometimes argued that the Browns Ferry accident discredits the Safety Study because the chances of it happening were supposedly "one in a trillion."[35] It is true that the Safety Study did not include in its analysis this specific scenario. But the predicted probability of a core melt, from *all* causes, is 1 in 20,000 per reactor-year. We have now had over 200 reactor-years of operation with similar reactors, and in that sense we were "due" for a near-miss with a one percent chance of a core melt. As indicated above, the Safety Study estimates that there was one chance in 300 of a core melt at Browns Ferry. Thus, we were not merely "due" for Browns Ferry; we were due for a near-miss which was three times closer.

Further misses are likely to occur in the future, and we will only keep a rational balance if we remember that even were the "miss" to turn into a "hit," and a core melt were to occur, this would not automatically be tantamount to a major human disaster. Again, in most cases of a core melt, there are expected to be no deaths.

On the positive side, it is to be noted that the Browns Ferry accident has stimulated improvements in design and operating procedures so as to avoid repetition of similar accidents. This illustrates the point we made earlier: experience with mistakes can lead to improved safety.

After all such arguments are made, critics of nuclear power often ask, "If nuclear power is so safe, how come the insurance companies won't insure it?" This question refers to the Price-Anderson Act, which sets the conditions for nuclear accident insurance. As with all insurance, by its nature, the Price-Anderson legislation reflects both confidence and questions. It can be interpreted, for what it is worth, as either a plus or minus for nuclear power safety. Our own impression is that the issues raised by Price-Anderson have more to do with the insurance industry than with nuclear safety, and are introduced into the nuclear discussion largely as a debating device. Nevertheless, in view of the attention the issue has received, we make brief comments on Price-Anderson in Appendix E.

A more visceral reason for public concern over possible reactor accidents is fear of genetic damage from radiation. Socially responsible people view with horror the possibility that we may be tampering with our gene pool and threatening the future of the human race. The basis for this concern is the demonstrated fact that high levels of radiation produce mutations in animals, and almost certainly would in man as well, given high exposures.

As in the case of radiation induced cancer, it is assumed that genetic effects will extrapolate in a linear manner from the levels where they have been measured to very low radiation levels. As discussed in Appendix C, this seems to be a conservative assumption. But even when one makes it, the calculated genetic consequences of possible accidents are small. For the worst case accident (once in a billion reactor years), the increase in genetic effects, over those occurring normally, would be only 0.1 percent for the country as a whole.[36] Further, the effects would appear most strongly in the first generation, and contrary to some popular impressions, succeeding generations would have a gradually decreasing burden from a single accident.[37] In short, the human race, exposed to radiation from its inception, is not genetically sensitive enough to radiation to be appreciably affected by anything the reactor program could do, even assuming the worst.

An indication of the exaggerated nature of conventional fear of genetic damage is the popular image of the consequences of Hiroshima and Nagasaki. All of us have seen, or think we have seen, pictures of deformed babies of parents who were bomb victims. However, there is no clear evidence of *any* genetic damage caused by the two bombs.

That statement concerning genetic bomb damage is so directly contrary to common belief that it is generally greeted by incredulity; in fact, such a reaction of disbelief was displayed by one of the authors when first confronted by the statement. Nevertheless, it is based on the outcome of careful investigations of the Hiroshima and Nagasaki populations. Not only was the conclusion reached in the joint Japanese-American Bomb Casualty Commission studies of survivors and their offspring; it has been confirmed in a re-examination of the evidence carried out by the Danish geneticist, Ove Frydenburg.[38]

It cannot be said that the bombs produced no genetic damage, but whatever damage was produced in the enormous explosions was too small to be identified positively, in view of the much greater amounts of genetic damage that arises, continually and naturally, in all populations.

In conclusion, no basis can be found for the contention that the danger of nuclear-reactor accidents poses a serious problem. In fact, the risks do not approach those of many other man-made and natural hazards, some of which arise from other forms of energy generation.

However, we do not expect that complete unanimity will be reached on the safety issue, any more than on other issues which excite great public interest.

Until a few years ago a group called the "Flat Earth Society" existed in England. While the justification for concern over reactor safety certainly is not to be compared to any justification for belief in a flat earth, it is not reasonable to take the absence of unanimity *by itself* as constituting a viable argument against the deployment of nuclear power.

REFERENCES

1. Nucleonics Week, p. 8, Nov. 16, 1975. The operating performance of most reactors in the Western World are summarized each month in Nucleonics Week.

2. Department of the Navy, Commander Naval Sea Systems Command, Wash., D.C., letter to authors from W. Wegner (June 19, 1975).

3. See Ref. 5 below, Main Report, p. 135.

4. *Reactor Safety Study (Draft)*, U.S. Atomic Energy Commission (Aug. 1974).

5. *Reactor Safety Study: An Assessment of Accident Risks in U.S. Commercial Nuclear Power Plants*, WASH-1400, U.S. Nuclear Regulatory Commission (October 1975). (National Technical Information Services, Springfield, VA.)

6. Ref. 5, Executive Summary, p. 8. Since this and other results of the Safety Study are at the center of much of the debate, exact page number references are included here and in the references which follow. These are difficult for readers interested in

the further details to locate because of the large size of the Safety Study. The Main Report, which includes the Executive Summary, is the most important. It may be purchased for $7.75.

7. Ibid., Executive Summary, p. 9.

8. Ibid., Main Report, p. 73.

9. Ibid., Main Report, p. 104.

10. Ibid., Appendix VI, p. 13-32.

11. Ibid., Main Report, p. 104.

12. Ibid., Main Report, p. 119.

13. Ibid., Executive Summary, p. 2.

14. Ibid., Main Report, p. 66.

15. Ibid., Main Report, p. 90.

16. Ibid., Executive Summary, p. 11.

17. Ibid., Main Report, p. 19.

18. Ian A. Forbes et al. *Nuclear Reactor Safety: An Evaluation of New Evidences,* Nuclear News, p. 32, Sept. 1971; D. F. Ford et al., *The Nuclear Fuel Cycle,* Union of Concerned Scientists, PO Box 289, M.I.T. Branch, Cambridge, Mass.

19. Ref. 5, Main Report, p. 140.

20. S. H. Day, Jr., *Scant Cause for Reassurance,* Bulletin of the Atomic Scientists, Oct. 1974; F. H. Schmidt, *Comments on Scant Cause for Reassurance,* Bulletin of the Atomic Scientists, Jan. 1975.

21. Press release. See, for example, *The National Observer,* Aug. 31, 1974.

22. Närförläggning av kärnkraftverk, Stockholm, 1974. See particularly p. 284. (Summary, in English.)

23. Ref. 5, Appendix XI.

24. Sierra Club Report, Sierra Club, 1050 Mills Tower, San Francisco, CA., Union of Concerned Scientists, 1208 Mass. Ave., Cambridge, Mass.

25. N. C. Rasmussen, Spectrum (Institute of Electrical and Electronic Engineers) p. 46, Aug. 1975.

26. J. M. Døderlein, *The Rasmussen Report Revisited,* Institute of Atomenergi, Kjeller, Norway.

27. Nucleonics Week, p. 3, Nov. 21, 1974, and p. 3, Dec. 12, 1974.

28. *Report to the APS by the Study Group on Light Water Reactor Safety,* Reviews of Modern Physics, Vol. 47, Supplement 1, (Summer 1975). Reviewed in *New York Times,* April 29, 1975.

29. Ref. 5, Main Report, p. 97.

30. Associated Press, May 19, 1975.

31. Ref. 5, Main Report, p. 74.

32. Nucleonics Week, Mar. 27, 1975.

33. *Fire in the Cable Spreading Area and Reactor Building on Mar. 22, 1975.* U.S. Nuclear Regulatory Commission #50-259/75-1 & #50-260/75-1 (July 25, 1975).

34. Ref. 5, Appendix XI, p. 3-55.

35. *The Case for Nuclear Safeguards,* Coalition for Safe Energy brochure, Seattle, Wash. 1976.

36. Ref. 5, Main Report, p. 141.

37. *The Effects on Populations of Exposure to Low Levels of Ionizing Radiation* (The BEIR Report), p. 43, National Academy of Sciences—National Research Council (Nov. 1972); Ref. 5, Appendix VI, p. 9-30.

38. Ove Frydenberg, Naturens Verden (Nature's World), October 1974, p. 305. In Danish.

BREEDER REACTORS—
PROSPECTS AND PROBLEMS
Chapter Fourteen

Most of the discussion in earlier chapters has centered on the debate over the commercial light water reactors (LWR's) of the type now in operation or projected. However, as shown in Chapter 5, the long-range potential of the earth's uranium resources can be utilized only if the fertile U-238 is converted into fissile Pu-239. This is done in breeder reactors.

The breeder reactor therefore becomes an absolute necessity within a relatively few decades, unless, of course, a happy technological breakthrough were to occur on either the solar-energy or the fusion-energy fronts. Though reserves of coal might last for 100 or more years, coal presents serious long range problems, as discussed in Chapter 17.

At present, the prospects for either solar or fusion energy are much less certain than for the breeder. It therefore appears crucial to forge ahead vigorously on breeder reactor development. Should its problems prove to become more, rather than less, tractible as development proceeds, the nation and world would be little worse off for the effort, since the development cost is not a genuine concern as compared with other expenditures for national security, or for that matter, with oil import costs.

The debate over breeder reactors revolves around two chief issues: safety and economics. Additionally, there are other less heated elements of the debate, some of which stem from minor misconceptions.

In what follows, we first consider the general status of the breeder; next we turn to a few misconceptions. To appreciate why the debate arises over safety it is necessary to discuss technical questions at greater depth than was required for the light water reactors. Some of the technical discussion reflects features of the challenge presented by breeders. Last, we discuss aspects of the economic debate. For both safety and economics an analogy can be drawn with the history of jet aircraft.

Contrary to much popular opinion, breeder reactors have been under development for many years. The first reactor ever to generate electricity was an experimental breeder.[1] Foreign countries have breeder programs, and some are more advanced than the US. Their concern can be partly appreciated when it is recalled that only the US and USSR have coal resources to sustain even constant energy needs for reasonable times.

France, with its reactor called Phenix, and the USSR appear to be most advanced,[2,11] although recent reports suggest the first large Russian breeder struck a snag.[3] England and West Germany (in cooperation with other countries) both are pursuing active prototype breeder programs.

France's Phenix produces a maximum of about 250 megawatts of electric power. During the first 11 months of 1975 it operated at an average of 66 percent capacity.[4] This compares favorably with light water reactors, and with coal fired plants, and is especially noteworthy because it was not brought on-line until late 1974.[5]

Presumably, Phenix is producing more Pu-239 than it consumes, but the true ratio will probably not be known until the fuel is removed for reprocessing.

Already France is designing a Super-Phenix, and reportedly has contracted to supply several to Iran. Similarly, larger breeders are being planned in the USSR and elsewhere.[2]

In the US a breeder a little larger than Phenix is being designed and is to be built on a site in Tennessee along the Clinch River.[2] The program has been under review, and attack, in assorted congressional hearings, environmental groups, administrative bodies, etc., etc.

Another large portion of the US breeder program consists of a test reactor, the "fast flux test facility," under construction at Hanford, Washington. Its purpose is only for research and development. It will not produce electric power.

A small prototype commercial breeder (60 MW(e)) called the Fermi reactor (after Enrico Fermi, one of the physicists most responsible for the first chain reaction achieved in 1942), was built near Detroit by the Detroit Edison Company. A metal vane placed, ironically, in the reactor vessel purely as a safety feature, broke loose and partially blocked a cooling passage. The core suffered a partial melt-down. No injuries occurred, and the reactor was ultimately repaired and operated successfully.[6] This accident occurred in 1966.

At no time was there significant danger to the public, but the Fermi reactor accident serves to prove, in some people's minds, that breeder reactors cannot be safe.

One possible course for the US might be to drop the breeder program entirely, and just wait for the French to do it. If their Super-Phenix is a great success, all we would have to do is buy it from France. Some people can find considerable merit in such a proposal. For example, since technicians and engineers would have to learn to run the Phenixes, they might have to learn French first, and that could be regarded as a cultural "spin-off" of the program.

Another school of thought runs as follows: If the French have done it, we certainly can also. The situation is, perhaps, similar to that which prevailed at the end of World War II, when the chief "secret" concerning the atomic bomb was revealed by just letting the world know "it could be done."

However, in the case of Phenix, the questions and motivations are more complex: What are the real costs? Is Phenix really safe? Is it actually breeding? Do we like the design? Might we not do better? What about commercial licensing problems?

Most of the issues of the nuclear debate are common to both light water and breeder reactors. The only real exception concerns the safety of the reactor itself, about which there is a diversity of opinion even among experts.

Although breeders obviously produce more plutonium than LWR's for a given electrical energy production, the LWR's produce about one third as much. Thus, the basic problem of handling the plutonium is essentially unchanged from that for the light water reactors (with plutonium recycle).

It is often said that breeders would produce "huge" amounts of plutonium, and the implication is that this material will somehow pile up and grow and grow. First, if one characterizes one ton of plutonium production per year per reactor as "huge," then what adjective have we left in our language to describe the thousands of tons of pollutants injected into the atmosphere per year by a coal-fired power plant? Second, unlike the products of the coal plant, the plutonium will be (indeed, must be) consumed by recycling.

Another misconception is that breeders "produce more fuel than they use." Were that true, we would have a kind of perpetual motion machine. All breeders actually do is make more fissile material than they use, at the expense of consuming "fertile" U-238, which is, of course, in the broad sense, a fuel.

We turn now to breeder safety, and begin with a short description of the essentials of how breeders work.[7]

In Chapter 5 we outlined the basic physical processes occurring in a mixture of U-238 and U-235. We recall that in a light water reactor the extra neutrons produced by neutron-induced fission of U-235 are first slowed down by the water, then most of them are reabsorbed either by U-235 to produce more fission, or by U-238. The latter material, which first becomes U-239, radioactively decays within a few days and becomes Pu-239.

The exact ratio of the two events, fission or U-238 absorption, depends in a complex way upon the inherent probability nature has established for these processes, as well as upon some factors man can control in his designs; as for example, the distance between fuel rods, and the percent enrichment of U-235.

As a further complication, nature alters the chance of either fission or absorption when the speed of the neutrons is changed. Taking all such complexities into account, it turns out that to increase the production of Pu-239 to the point where it is equal to or greater than the loss of U-235, it is necessary to rely on faster neutrons (by using very little moderator), and to increase the concentration, or enrichment, of the U-235.

It is also necessary to conserve neutrons as carefully as possible. Light water is a poor substance for this purpose because it both tends to absorb neutrons, and to slow them down too much.

To make a reactor breed, a coolant with a higher atomic weight, and one which does not absorb too many neutrons, is convenient. Also, a liquid is desirable in order to efficiently carry off the heat to make steam. For these and other reasons liquid sodium has been most favored.

The Pu-239 which is bred during the first cycle can be used in newly fabricated fuel to replace the original U-235. Thus, as time goes on, and fuel is again recycled, the world might eventually deplete the U-235 supply, but Pu-239 will take its place. This ultimate state might be referred to as the plutonium energy economy.

The breeding ratio is the amount of plutonium bred compared with the original amount in the fuel. Nature smiled on us in another way: The breeding ratio turns out to be greater for a Pu-239 and U-238 mixture than it is for a U-235 and U-238 mixture, and thus once started breeding gets easier, not harder. If a breeding ratio of (say) 1.1 can be achieved over the whole fuel cycle, then the Pu-239 "doubling time" (as for a bank account at 10 percent interest) would be seven years.

Of course, if society has enough light water reactors, the plutonium needed to start the breeders is already available. Conversely, any excess plutonium from the first breeders can be used in the light water reactors, or in new breeders.

In the safety area, nature was kind to us for the slow neutron, regular light water reactor. She provided decreased "reactivity" (the quantitative extent to which the reactor tends to be critical) as the temperature rises and produces boiling of the moderator. Also she caused many of the neutrons from fission to be "delayed." These factors tend to facilitate the control of the reactor and thus provide inherent safety features.

In contrast, under the conditions needed to make a reactor breed, nature was not so kind. At the higher neutron speeds there are fewer delayed neutrons from fission. To stabilize the reactor power level, the mechanically driven control rods must therefore respond more quickly. The result is a decrease in controllability, and an increased potential for sudden overheating.

An analogy might be made with the difficulty of steering an automobile whose front wheels are badly misaligned, so that every time you adjusted the direction of motion the car "over-shot" and steered too far the other way just because you could not react fast enough.

Fortunately, after being so unkind as described above, nature gave us a mitigating effect, the net result of which causes the reactivity to decrease if the temperature rises. It is called "Doppler broadening." It is a complicated but rather interesting effect, and we will describe it because it is so important for the safety issue.

The Doppler effect in sound is a familiar phenomenon. If an automobile sounds its horn while passing you, you hear a higher pitch sound during its approach than while it is receding. The sound waves tend to scrunch together when the automobile is approaching, and to spread apart when receding, causing a change in frequency or pitch.

The atoms of U-238 and of U-235 have affinities for absorbing neutrons only if the neutrons have certain specific speeds. There are numerous "right" speeds for both isotopes, but they are a different set for each. It is as if the uranium atom is "tuned," and the speeds have to match one of the tuned channels.

When the temperature of the fuel increases, the atoms of uranium move about more rapidly in random directions. This tends to increase the range of neutron speeds over which the right tuning is hit to allow a U-238 nucleus to absorb neutrons, or a U-235 (or Pu-239) nucleus to fission.

The reason the range of effective neutron speeds increases is because only the *relative* speed between the absorbing atom and the neutrons is important. Thus, the neutrons are more readily absorbed by U-238 and by U-235 (or Pu-239) as the fuel temperature goes up.

It is as if many violins were putting out steady musical notes, each different. If your ear were tuned to hear only, say, 10 precise tones the chances are you would miss hearing most of the violins. But if you moved around fast enough, darting back and forth in all directions, you would "detune" your ear by the Doppler effect and thus end up hearing more of the violins. The violins in this example are the neutrons, the ears are the uranium nuclei, and the darting about is the effect of heating up the uranium.

For this reason, both the chance for fission in U-235 (or Pu-239) and the chance for absorption in U-238 are enhanced by an increase in temperature. Were the former rise greater than the latter, the reactivity would "run-away." But here nature turned out to be kind! The Doppler effect change is larger for the U-238 component than for the U-235 (or Pu-239) component. Overall the fission loses out with increasing temperature and the reactor corrects itself.

The Doppler broadening effect in breeder reactors was first predicted theoretically. It has been tested experimentally (a tricky business in a real reactor) and found indeed to be correct. It thus acts like a temperature sensitive "damper" and mitigates the bad effects of increased temperature.

The material favored by most experts for the coolant is liquid sodium. Unlike some other possible substances for moderators, such as helium gas or heavy water, the earth's supply of sodium is neither limited nor is it difficult to extract. A breeder reactor utilizing sodium is referred to as a liquid metal fast breeder reactor, or an LMFBR. The "fast" in the title refers to the speed of the neutrons.

Sodium metal becomes liquid (when under atmospheric pressure) at a temperature only about 4°F below the boiling point of water. It is thus very easy to liquify. Moreover, it does not boil until it is heated to over 1600°F.

Therefore, a sodium-cooled reactor can operate at atmospheric pressure—a strong safety advantage for the LMFBR. Liquid sodium also has a high capacity for removing heat, and thus is an effective coolant.

On the other hand, liquid sodium has very disagreeable chemical characteristics. It reacts wildly if it comes in contact with water. It also becomes very radioactive when bombarded with neutrons. Fortunately, this radioactivity is short-lived: it dies away with a 15-hour half-life. In 300 hours it would be at one-millionth of its original level. A very small fraction is longer lived.

The LMFBR design has a closed loop of sodium coolant which heats a second closed loop of liquid sodium. The latter then exchanges its heat with water to produce steam, which then drives the turbines. The "double loop" system makes it nearly impossible for radioactive sodium to come in contact with water.

Just as in the light water reactors, the potential for accidents stems from the possibility of insufficient cooling. However, unlike the case for water in a LWR, if the sodium coolant boils and produces "voids" or if the coolant were lost from the reactor vessel, the fuel first tends to become more, rather than less, reactive. As the temperature rises, the Doppler effect takes over and decreases the reactivity. However, if the loss-of-coolant event occurred fast enough, the core could blow itself apart. The result is referred to as a "hypothetical disassembly accident." The reactor vessel is designed so as to withstand the force of the disassembly. It is "hypothetical" because it is only conceptually possible.

If the hypothetical disassembly accident actually occurred, then it is conceivable that the fuel could subsequently fall to the bottom of the reactor vessel, where it might become critical all over again despite the presence of absorbing material from the control rods.

This possibility exists for the breeder, but not for the light water reactor, because the breeder fuel is much richer, about 15 percent, in fissile material.

The destructive possibilities for both the above-hypothesized accidents are greatly decreased by the Doppler effect. The calculations have been supported by actual "destructive" core tests in small breeder reactors.

Thus, to summarize, there are some reasons for believing the accident potential of the LMFBR is greater than for the LWR; conversely, there are some factors in favor of the LMFBR. The situation at the present time might be compared with the time in 1957 when WASH-740 (see Chapter 13) first assessed LWR potential accidents. In short, no in-depth study of LMFBR's on quite the scale of the Rasmussen Reactor Safety Study has yet been made. Indeed, it would be difficult to do so before further design and development work proceeds on the LMFBR. It therefore is not suprising that there is some diversity of opinion among the experts as to the actual accident risk for breeders.

One additional advantage of the LMFBR is that it can produce steam at a higher temperature than can a LWR. Consequently, the efficiency with which electrical energy is produced is greater—40 plus percent, compared with about 32 percent. The waste heat produced by the LMFBR is therefore less, and it consumes less fuel for a given electrical output.

There are other types of breeder reactors which have future possibilities. For example, a slow neutron breeder may be feasible which would utilize thorium and breed U-233. The world's supply of thorium is even greater than its supply of uranium, and together they represent hundreds of thousands of years of ample energy.

The rising tide of debate over the breeder program has spread to the issue of rising cost projections.[8] It is the capital investment costs which are important. Just as for solar energy, breeder fuel costs are negligible as compared with capital costs. Operating costs are another matter, however, and comparisons between solar and the breeder are difficult to make.

A recent study[9] by experts at the Hanford Engineering Development Laboratory in Washington State presents calculations which show that if the LMFBR is introduced by 1987 and expanded thereafter, the cost of electrical energy will be reduced by a total of 72 billion dollars over some 50 years, or an average of 1.5 billion per year, as measured in constant 1974 dollars. The comparison is measured against deployment of only light water reactors.

Other analyses[10] are as unoptimistic in their economic projections as the above is optimistic. The debate is exacerbated by steadily rising projected costs for the prototype Clinch River Breeder Reactor. Much of the projected rise is evidently due to more and more added safety features. To give a comparison, the French breeder Superphenix is expected to cost about $1,000 per installed kilowatt,[11] which is about 30 percent higher than a LWR at 1975 prices.

The present breeder reactor situation is somewhat analogous with the historical development of commercial jet aircraft.

Jet airplanes had their origin in the successful jet fighters built in Germany near the end of World War II. The first commercial jet was the Comet, built in England in the early 1950's. The Comet had a very unfortunate accident record, and was soon abandoned. Moreover, it also was apparently unduly expensive and unable to fly very far.

In technical circles in Seattle (the home of Boeing) evening conversation often dwelt on questions such as whether jet airplanes could ever be safe, be economical, and have sufficient range.

The Boeing Company built a commercial prototype—the famous 707—modeled after military aircraft. It was a financial gamble. It was flown around the world for a year to prove its reliability and economic viability. That was in the late 1950's. Today, the whole world is aware of the great success of the Boeing 707, and of the models which followed.

It is just possible that the Fermi breeder reactor is the "Comet," but unlike the Comet, no one was hurt by the Fermi reactor accident. It remains to be seen if the Clinch River Reactor becomes the 707. But when gauged by the relative sizes of the organizations involved and by the magnitude of the stakes, the two technologies are not very different.

We close with the point made earlier. Although we do not know all the economic and safety problems involved in the breeder, these can only be understood if work on prototype breeders is carried out. Breeder technology, despite its incompleteness, is more advanced than solar energy or fusion technology. All three should be simultaneously pursued, but of all the avenues the breeder looks the most promising. It is for this reason that France, the USSR, and other countries of Western Europe are so determined to develop breeders. For the United States, the breeder is a bet which probably will pay off. It is like betting on a horse where the payoff is 100 to 1, and the chance of the horse winning is better than even money.

REFERENCES

1. This event occurred in 1952. It is described in dramatic terms in *Atomic Shield*, p. 497, R. G. Hewlett and F. Duncan, Penn. State Univ. Press (1972). Republished as WASH 1214, Vol. II, U.S. Atomic Energy Commission (1972).

2. ERDA-1, UC-79. U.S. Energy Research and Development Administration (Jan. 1975).

3. Ibid., attachment 7, p. 7.

4. Nucleonics Week, Dec. 18, 1975.

5. Although the first full power was achieved in March 1974, Phenix was not operated steadily on the French electrical grid until later that year. See monthly listings of most of the world's reactors, Nucleonics Week.

6. *Fuel-Melting Incident at the Fermi Reactor on Oct. 5, 1966*, R. L. Scott, Nuclear Safety *12*, 123 (April 1971).

7. Additional details are well described in *Nuclear Science and Society*, Bernard L. Cohen, Anchor Press/Doubleday (1974).

8. For example, David Burnam, New York Times, Nov. 16, 1975.

9. *The Incentive for the Liquid Metal Fast Breeder Reactor—A Revised Economic Analysis*, R. P. Omberg, R. W. Hardie, and J. H. Chamberlin (Oct. 1975.)

10. For example see *The Liquid Metal Fast Breeder Reactor—An Environmental and Economic Critique*, Thomas B. Cochran, Resources for the Future, Inc. (1974).

11. *European Breeders*, W. D. Metz, Science, *190*, p. 1279 (Dec. 26, 1975) and *191*, p. 368 (Jan. 30, 1976).

SAFEGUARDS AND SECURITY
Chapter Fifteen

A particularly frightening potential danger arising from nuclear power has been brought to public attention largely through the efforts of Theodore Taylor, formerly an AEC bomb designer, and later engaged in independent research and consulting on nuclear safeguards.

Taylor's career and the problems that concerned him were detailed in a series of articles in the *New Yorker* magazine and in a book based on the articles.[1,2] In addition, the problem is studied in somewhat more technical terms in a book by Willrich and Taylor.[3] Mason Willrich is a law professor formerly involved in issues of nuclear-arms control on the international level.

The thesis they present is straightforward: Nuclear bombs require a relatively small amount of fissionable material, such as U-235 or Pu-239; 10 kilograms or possibly less is sufficient for one bomb. Much greater amounts of such material are produced in the operation of nuclear power plants and in the fuel-enrichment plants. One can envisage a terrorist group stealing such material, either surreptitiously in small amounts over an extended period of time, or in a single attack on, say, a truck transporting nuclear fuel. According to Willrich and Taylor, it would not be difficult for them to fashion the material into a bomb.[5]

The theft of nuclear material could provide material for bombs only at certain stages in the fuel cycles. Regardless of reactor type, an attack on a reactor or on on-site storage facilities or theft of spent fuel during shipment from the reactor does not present problems. The reason is that the chemical separation of the plutonium (or of U-233 for reactors operating on the thorium cycle) from the highly radioactive fission products would be impossible in an operation on the scale one could conceive for any plausible terrorist organization.

However, the would-be terrorists would find a much more attractive target after the spent fuel has been reprocessed and the plutonium or U-233 separated, particularly during transport of the reprocessed material to fuel-fabrication plants or while fuel elements are under shipment back to reactors.

It is a matter of some dispute as to how easy it would be for any group to make a bomb out of such stolen material.[6] It is probable that Taylor could succeed himself, but as to the terrorists he envisages, it may be like saying a math problem can be solved by any high-school student with the brains of Einstein.

Nuclear explosions occur when a super-critical mass of fissile material is created suddenly and a burst of neutrons is generated to trigger the chain reaction. The simplest bomb consists initially of two pieces of nearly pure fissile material which are brought together very rapidly. At the instant of supercriticality, a modulated source of neutrons starts the reaction. The source can consist of a radioactive material emitting alpha particles which is brought together with beryllium metal at the instant of maximum criticality. The chance of premature explosion, or "fizzle," always exists, because neutrons are present at all times, even from cosmic rays. That's the reason the pieces must come together rapidly. The reaction generates gigantic quantities of heat and radiant energy. It proceeds until a portion of the fissile material is consumed and the entire mass flies apart in a violent manner.

Such a bomb as this probably could be made by a small group of moderately technically oriented terrorists. Even so, it might be a fair-sized project requiring a substantial array of facilities, at least as judged by the fact that the first bomb of this kind actually weighed about 10,000 pounds.[7]

J. Carson Mark observed that the simple-type bomb described in the preceding paragraphs would not work, even in a "crude" form, with plutonium from reactors as the fissile material.[8] The reason is that this plutonium is not pure enough.

In an earlier chapter, we described the physics of how neutrons could be absorbed by U-238 to produce first U-239, which then subsequently decays to Pu-239. In turn, once the Pu-239 is present in nuclear fuel, it can absorb neutrons and become Pu-240. Alternatively, in a fixed ratio, it can produce a fission reaction and thereby contribute to the energy generated. However, the Pu-240 is consumed less rapidly by fission, and thus tends to build up in concentration. After a year or more in the reactor—the typical time fuel remains in the reactor is three years—the fuel elements contain a considerable fraction of Pu-240. This isotopic material cannot be separated by chemical techniques from Pu-239. To perform such a separation would require a plant similar to the great diffusion plants used to separate U-235 from natural uranium. Thus, new fuel fabricated from reactor-grade plutonium is "contaminated" with Pu-240.

Fortunately, the Pu-240 has no particularly deleterious effects on subsequent reactor operation when plutonium is recycled. However, it possesses a property which greatly reduces its usefulness as bomb material. Although it mainly decays like ordinary radioactive substances, it also decays partly by spontaneous fission. In doing so, it produces extra neutrons in a manner entirely analogous to induced fission of U-235 and Pu-239. Thus a very large number of neutrons are continually emitted from a sample of reactor-produced plutonium.

As already indicated, to make a bomb it is necessary first to produce a super-critical mass and second to provide a burst of neutrons to ignite the chain reaction at the instant of supercriticality. In the presence of a bath of spontaneously generated neutrons, it is difficult to create a super-critical mass just during an instant in which no neutrons are present to cause a fizzle. Those neutrons are distributed statistically in time with no long waits between them.

The discovery of spontaneous fission of Pu-240 in the summer of 1944 during the Second World War almost caused the cancellation of the entire plutonium portion of the Manhattan Atomic Bomb Project. A description of the consternation it produced is found in "The New World."[9] In fact, any potential bombmaker would be immediately discouraged upon reading of the great troubles experienced at the Los Alamos bomb laboratory. The successful solution to the problem required a much more complex design utilizing the implosion method, but it was still necessary to limit the amount of Pu-240 permitted to build up as a contaminant in the fuel elements of the wartime reactors at Hanford.

As far as we know, no publication has stated the actual Pu-240 percentage in what is referred to by Willrich and Taylor[3] as weapons-grade material. However, one can infer it might be 8.6 percent or less.[10]

It is undoubtedly true that the deleterious effects of Pu-240 can be circumvented with more recent bomb designs. However, we have never learned the secret, although we have followed the open literature many years. That it can be done is made clear by J. Carson Mark,[8] but only during discussion assessing the future capabilities of nations not now possessing nuclear weapons.

We are of the opinion that it must be extremely difficult, requiring very sophisticated knowledge and skill, as indicated in the following discussion.

The actual amount of Pu-240 contamination, compared with Pu-239, in reactor-produced plutonium is quite large, about 20 to 35 percent for our present reactors.[4] With these values, together with the half-life for spontaneous fission in Pu-240,[11] it is easy to calculate the average time interval between neutron emissions and show that it must be very difficult to make a bomb using reactor plutonium.

An analogy might be made with the silicon transistor (it is almost all sand), about which everything is published. But no one person or any single group, for that matter, could make one without a great deal of study and technical expertise, backed by a highly sophisticated scientific technology.

India recently exploded a nuclear "device." Mrs. Indira Gandhi, the prime minister, stated at the time she was very proud of India's scientists for this *difficult achievement,* which took so long to accomplish. Why was it so difficult for India's scientists if it is supposed to be so easy for terrorists? Or, as Willrich and Taylor insist, why is it so simple for ". . . a few persons, possibly even one person working alone . . ." to fabricate a bomb?[3]

To us, the evidence on the "ease" of making a bomb is not convincing. On the other hand, Willrich and Taylor qualify their claims by referring to a "crude" bomb, one which would be about 1/20th the energy of the original atomic bomb which was equivalent to 20,000 tons of TNT. Apparently, a test of whether such a bomb, fashioned by unsophisticated methods within the grasp of conceivable terrorist groups and using reactor plutonium, would actually work has never been made, at least not to Taylor's knowledge. [12]

Perhaps in an effort to answer that question in a quantitative manner, the AEC (now ERDA and the NRC) is reported to have instituted a new study. [13] Were the probability of success shown to be significant, one solution to the safeguarding of fuel shipments has already been suggested by Alvin Weinberg. [14] It is to contaminate the plutonium fuel rods with fission products prior to shipment to reactors, thus rendering them essentially untouchable. The same scheme could safeguard highly enriched U-235 fuel, which is used in high-temperature, gas-cooled reactors operating on the thorium or U-233 cycle.

J. Carson Mark, formerly head of the theoretical division at Los Alamos, believes it would require six highly skilled people to make a "crude" bomb: [15]

> "If one thinks of a small group wanting to build a bomb, and if one supposes that the primary requirement is that it give a 'nuclear yield' (as to say, for example, 'the yield must be at least so much; but it is all right if it should turn out to be a few times larger'), then I think that such a device could be designed and built by a group of something like six well-educated people, having competence in as many different fields. As a possible listing of these, one could consider: A chemist or chemical engineer; a nuclear or theoretical physicist; someone able to formulate and carry out complicated calculations, probably requiring the use of a digital computer, on neutronic and hydrodynamic problems; a person familiar with explosives; similarly for electronics; and a mechanically-skilled individual. Among the above (possibly the chemist or physicist) should be one able to attend to the practical problems of health physics which would arise. Clearly, depending on the breadth of experience and competence of the particular individuals involved, the fields of specialization, and even the number of persons, could be varied, so long as areas such as those indicated were covered."

It is very hard to envisage a terrorist organization capable of marshalling such an array of educated persons in a clandestine manner, replete with rather sophisticated laboratory facilities. To design effectively one would need to know the percentage of Pu-239 relative to other plutonium isotopes present in the stolen material. It is likely the composition would differ from one fuel sample to the next, perhaps in a random manner, a fact which would render the task of the terrorists even more difficult. In addition, plutonium is a dangerous radiologically toxic material, even discounting the "hot-particle" theory.

Another assessment of the terrorist problem is contained in a Swedish report. [16] Although concerns over the various nuclear problems are most intense in the U.S., they are not unique to us. Sweden derives about 70 percent more electricity per capita from nuclear power than does the United States and the question of nuclear terrorism has appropriately attracted attention there, as well. The Swedish report on this subject concludes:

> For an illegal organization, a supply of qualified personnel and (the obtaining of) a supply of fissionable material would entail difficulties. Above all, the recruiting of some 50 men among whom 10 or so are qualified specialists and several years of activity in secrecy could probably not be carried out. These problems combined with the high costs for such an organization make it improbable that the organization would engage in the production of nuclear explosives. A single madman (a terrorist?) completely lacks the possibilities for producing a nuclear explosive.

From all this discussion, it is apparent that terrorists would have a difficult, perhaps nearly impossible, job in fashioning a bomb from stolen plutonium. It is a job for a band of very highly trained people. Nevertheless, prudence suggests that we assume that the terrorists might achieve the requisite training or enlist the help of reasonably sophisticated nuclear scientists. Even if their success is not assured, just a small prospect of success might serve the ultimate purposes of the terrorist group, especially if the possible effects of their acts are overestimated by the public.

Beyond any damage the terrorists could do with a plutonium bomb, there is also concern about the effects of plutonium dispersal by a crude nuclear device, or just by a chemical explosion. As discussed in Chapter 11, the consequences would not be as drastic as sometimes portrayed. It might cause about 30 cancer deaths per pound dispersed in a typical city. Dispersal of 20 pounds would not wipe out the city, but it would represent a major disaster. In selected sites, such as a football stadium, the death rate could be much higher.

For these reasons it is essential that attention be paid to the prevention of plutonium theft. If safeguards could be made essentially impregnable, it is obvious that questions of whether it is easy or isn't easy to fabricate a bomb or whether plutonium is or is not extraordinarily toxic quickly become moot. Thus, Taylor's asserted goal in bringing these matters to public attention—and simultaneously subjecting himself to criticism for giving ideas and possibly even guidance to potential saboteurs—was to jog the AEC into taking adequate steps to reduce the chance of such theft.

Willrich and Taylor[3] suggested various ways to provide adequate safeguards for nuclear materials in each step of the fuel cycle. In addition, a symposium[18] devoted to the subject developed other ideas demonstrating clearly that imaginative methods for safeguards exist which easily match the imaginations of those critics who endow terrorists with capabilities beyond demonstrated reasonableness.

It is important to remember that the actual amount of material that must be safeguarded is very small, simply because the energy content is so enormous. Remember that a kilogram of plutonium is equivalent to more than 2,000 tons of coal. One thousand large breeder reactors would produce about a million kilograms of plutonium a year, which sounds like a lot to safeguard. However, that annual amount, together with containers, can be transported from a reprocessing plant to a fuel-fabrication plant by about 1,000 medium-sized trucks, or by about three trucks a day!

Even if each truck were accompanied by a convoy of armed guards, plus a few helicopters, the resultant "police state" aspects, feared by some critics, would scarcely be visible to any individual citizen. In sharp contrast, a million or more Americans accept as routine each day a much more visible reflection of a "police state" when they willingly submit to metal-detector and other tests to fly in a passenger plane or simply to watch a friend prepare to do so.

Although few people are aware of the presence of such safeguards, we already have a well developed system for the protection of nuclear weapons shipments.[17] A hijacked weapons transport truck is automatically disabled and cannot be moved. An automatic alarm is sent by radio to military authorities; entry into the truck is almost impossible. Similar measures could be adopted immediately for the transport of civilian nuclear materials.

In speaking of costs, Willrich and Taylor suggest that when U.S. nuclear-power utilization moves into high gear in the 1980's, the price of a safeguards program will be about $80 million a year. That may appear to be excessively high at first glance, but it would be less than 1 percent of the total cost of the nuclear-generated electricity. A 1 percent increase in individual electric bills would be minor compared to the benefits received from a successful safeguards program, even if it is only to put people's minds at rest.

Having now acknowledged the problem and the need for vigorous steps, thanks in part to the prodding of Taylor's campaign, the federal government has already developed a greatly improved safeguards program.[18,19] Both Willrich[20] and Taylor[21] have stated they consider the planned new safeguards to be adequate.

Despite all the deterrents and safeguards, one other sophisticated, "long term" danger might be considered: theft of the small percentage of plutonium that resides in stored wastes. It could be argued by some that our generations are leaving to posterity a growing possibility of terrorist acts, and that guarding these radioactive leftovers will be an impossible long-term task. As already pointed out in an earlier chapter, these wastes are not particularly dangerous to handle after some 500 or 600 years, so it might be easy chemically to separate the plutonium. But even if the wastes were accessible, an added deterrent is present: a significant quantity of Curium-244. It is an isotope that decays into Pu-240 with a half-life of 18 years; the result is that the fraction of Pu-240 present in the waste more than doubles within 30 years,[22] thus rendering the plutonium still less useful as bomb material.

With all these barriers, we submit that the clever terrorist-saboteur will find many other targets which are more attractive than attempted theft of plutonium in the U.S. For example: obtaining fissionable nuclear materials from friendly countries which have their own national sources of plutonium and enriched uranium, or obtaining chemical or biological weapons either by home manufacture or theft.

Even without such "exotic" weapons as those noted, terrorists can disrupt society through the use of ordinary bombs, guns, and kidnapping. With such alternatives in mind, there is little reason to believe society will be much more vulnerable with a civilian nuclear-power program than without it.

Prudence suggests reasonable safeguarding of nuclear material is a wise course. The much more difficult program of safeguarding thousands of airplane flights against hijacking has been very successful in the U.S., although it has not been able to insure *absolute* safety. In comparison the problem of providing a very high degree of security for the limited number of nuclear facilities and nuclear shipments should not be difficult.

Despite the formidable obstacles, a terrorist group bent on trying to produce or obtain a nuclear bomb will not be dependent on stealing plutonium produced in the U.S. Terrorism is international in scope, and so is nuclear energy. Those are the realistic facts.

Western Europe and Russia are committed to large-scale nuclear-energy use, and it is spreading worldwide, independent of any policy adopted by the U.S. For example, the French Ministry for Industry has announced Iran will help finance a diffusion plant to be built in France that will supply enriched uranium to Western European nuclear-power plants.[23] Despite being presently awash in oil, Iran has already ordered many nuclear-power plants, and will receive enriched uranium to use in them in return for its investment in the diffusion plant.

Similarly, Brazil is scheduled to obtain from Germany not only nuclear reactors, but also a plant to enrich uranium and a reprocessing plant. This will give Brazil a complete range of nuclear facilities, and can offer terrorists there targets similar to those which exist in the U.S. We recently succeeded in discouraging the purchase of a pilot plutonium reprocessing plant by South Korea from France,[24] but there are now few countries in the world which are still as much under U.S. influence as South Korea.

There is nothing inherently wrong or unnatural in the determination of Iran, Brazil, South Korea, and most of the rest of the world to obtain the advantages of nuclear energy. But their actions illustrate the impossibility of our gaining freedom from the threat of nuclear terrorism by curtailing our domestic nuclear power program. If a group is bent on nuclear terrorism in the U.S., it can search the world for the easiest place for theft, and then smuggle in the materials.

The limits of our influence on the spreading of these facilities is indicated by our experience with Brazil. When we declined to provide Brazil with the array of plants she wanted, Brazil turned to Germany. Within Germany, in turn, there have been some environmental protests against local nuclear reactors. But any protests which Germans may have made against the deal with Brazil have so far been so muted as to have been inaudible.

In this situation, the best the U.S. can do is to make nuclear theft as difficult as possible for domestic terrorists, and, by example and persuasion, attempt to induce the rest of the world to adopt tight safeguards measures.

REFERENCES

1. John McPhee, The New Yorker (Dec. 3, 10, 17, 1973).

2. John McPhee, *The Curve of Binding Energy,* Farrar, Strauss and Giroux, New York (1974).

3. Mason Willrich and Theodore B. Taylor, *Nuclear Theft: Risks and Safeguards,* Ballinger Publishing Co., Cambridge, Mass. (1974).

4. Theodore B. Taylor, *Nuclear Safeguards,* Annual Review of Nuclear Science 25, 407 (1975).

5. Ibid., Ch. 2.

6. Ref. 3, p. 21: Willrich and Taylor concede that other experts "will strongly disagree" with their contention that it is easy to make a bomb.

7. J. Carson Mark, *Nuclear Weapons Technology,* Impact of New Technologies on the Arms Race, M.I.T. Press, Cambridge, Mass. (1971).

8. Ibid., p. 139.

9. R. G. Hewlett and D. E. Anderson, *The New World,* pp. 247-251. (WASH-1214 Vol. I) (1972). Also, a description of the difficulties encountered in the development of the plutonium bomb is given by Stephane Groueff, *Manhattan Project—The Untold Story of the Making of the Atomic Bomb,* Little, Brown & Co., Boston (1967), pp. 321-330.

10. See Leonard Beaton, *Must the Bomb Spread?* Penguin Books (1966), p. 97. Up to July 1, 1962, the AEC paid more for plutonium with less than 8.6 percent Pu-240 contaminant. After that date, the price was independent of purity. Beaton interprets this to mean that bomb technology then was advanced to the point where the Pu-240 was unimportant. If so, then it must have been a long, hard road for the U.S. to achieve this level of sophistication: 1962-1945 equals 17 years. On the other hand, one might interpret this as evidence that stockpiles of good plutonium were sufficient. Whichever interpretation one accepts, terrorists would find that the high concentration of Pu-240 in reactor-grade plutonium presents a difficult barrier.

11. R. Vandenbosch and J. R. Huizenga, *Nuclear Fission,* Academic Press, (1973), p. 46.

12. Theodore B. Taylor, remark at American Physical Society Symposium, Washington, D.C. (Apr. 1975); see Ref. 18.

13. Nucleonics Week, July 4, 1974, p. 9.

14. A. Weinberg, Project Independence Hearings, Seattle, Wash., Sept. 5, 1974.

15. J. Carson Mark, private communication.

16. Nils Glyden and Lennart W. Holm, *Risks of Nuclear Explosives Production in Secret,* Swedish Report FOA4, C4567-T3, March, 1974, Förvarets Forskninganstalt, 10450 Stockholm 80. Translated report, ERDA-tr-45, NTIS, U.S. Dept. of Commerce.

17. Testimony before the subcommittee of the Joint House-Senate Committee on Atomic Energy, Review of the National Breeder Program, statement of Orval Jones, Director, Nuclear Security Systems, Sandia Laboratories, June 18, 1975.

18. *Safeguarding Nuclear Materials,* Symposium of the Forum on Physics and Society, Bulletin of the American Physical Society, *20*, 700 (Apr. 1975).

19. Reported in Nucleonics Week, June 6, 1974.

20. Mason Willrich, in a public lecture, Univ. of Wash. (May 20, 1974).

21. Before the Senate Committee on Banking, July 24, 1974, Federation of American Scientists—Public Interest Report, Vol. 3, No. 1 (Jan. 1975).

22. This result follows, for example, from *High Level Radioactive Waste Management Alternatives*, BNWL-1900, Battelle Pacific Northwest Laboratories (May, 1974), Vol. I, Appendix 2C.

23. International Herald-Tribune, Jan. 3, 1975.

24. New York Times, Feb. 1, 1976.

NUCLEAR PROLIFERATION
Chapter Sixteen

The close association between nuclear reactors and nuclear weapons suggests immediately that nations possessing nuclear-power plants automatically achieve nuclear-weapons capability. It is not quite so simple as it may seem. Nevertheless, the possibility presents a serious international problem requiring analysis and study far beyond the limited scope of this book. The purpose of this chapter is to outline briefly the major issues.

Light water reactors use uranium enriched in the isotope U-235 to a level far below that required to produce a rapid explosive chain reaction; consequently, LWR fuel cannot be used for bombs. The enrichment plants are complex and expensive and their creation is probably beyond the capability of any but large and affluent nations.

Thus, even small nations possessing adequate deposits of uranium ore would be unlikely to achieve nuclear-weapons capability using the relatively simple kind of bombs that can be devised with U-235 (see Chapter 15). The only way they can obtain fuel for their reactors is through international arrangements.

On the other hand, the spent fuel elements contain Pu-239, which after chemical separation and appropriate fabrication, can be used to make nuclear explosives or weapons, as described in Chapter 15.

It would appear, then, that any country—or a small group of very skilled people, for that matter—possessing even a small-sized nuclear reactor might possibly be able to make nuclear weapons. Although it is not economically feasible to use small nuclear reactors to make electric power, such reactors are useful for research and training, and many of them exist throughout the world. Numerous countries were aided in the acquisition of these reactors through the early efforts of the U.S. "Atoms for Peace" program.

Although we have referred to these reactors as "small," it is not intended to imply they are nice table top nuclear burners. They are actually rather large and complicated devices requiring large buildings, and, being so, they could hardly be built and concealed by a private group. They are small in contrast to modern commercial plants or those producing weapons.

A nation can proceed on its own and develop isotope-separation facilities and/or nuclear reactors, plus explosives and weapons, provided it has the necessary scientific and technical expertise, as well as the necessary industrial and natural resources. China and France are examples of countries which proceeded along that route. It is important to note that both nations deemed it necessary to *test* their bombs, presumably to find out if they worked. The tests are relatively easy to detect, worldwide, by means of seismic or other measurements.

Clearly, the only way to prevent nuclear proliferation of that kind is through international agreements. India recently produced a nuclear explosive device, essentially by the "home-made" route, but a portion of the technical and material assistance for the reactor was supplied by Canada, though presumably not the explosion technology.[1,2]

What deterrent, if any, exists then for preventing *any* country having, say, a light water reactor from extracting plutonium from the fuel rods and building bombs with it? Two barriers lie between the power plant and the bomb: a technical barrier and an international safeguards barrier.

In Chapter 15 the probable difficulties of producing bombs from reactor plutonium containing a large percentage of Pu-240 were outlined. If the plutonium is relatively free of the isotope Pu-240, it is certainly much simpler. But even so, to judge by newspaper reports,[3] it was not deemed easy for India to make a nuclear explosive, and no reason can be found to believe the Indians did not use relatively pure Pu-239. The next step—taking the process from an explosive device to a real bomb—is certainly not easy, but it is probably less difficult than that required to produce the explosive device itself.

The international safeguards barrier is provided by the rules and regulations of the International Atomic Energy Agency (IAEA), with headquarters in Vienna. In addition, a nation like the U.S. might impose its own safeguards as a condition for technical and other assistance.[3] Obviously only nations which agree to these regulations and live up to them are deterred by these safeguards. In India's case, the international safeguards apparently applied to her power reactors but not to a research reactor that produced the Pu-239. [1,2]

The effectiveness of the two types of barriers is open to debate, and considerable grounds exist for concern over the proliferation of weapons. It is doubtful, however, that these international questions have any significant implications for the U.S. domestic nuclear-power policy.

Obviously, U.S. construction of nuclear-power plants has nothing to do with U.S. weapons proliferation; the American nuclear arsenal's destructive capabilities are already awesome, and work continues on improving the delivery systems. Similarly, the curtailment of the U.S. nuclear energy program will in no way influence Russia, China, or France to diminish their own weapons programs.

As far as proliferation in non-nuclear countries is concerned, it would be possible for the U.S. to adopt a policy providing for a rapid buildup of its own nuclear-power plants without helping other nations to build their own plants. Although this is not the present policy and may never be, it would not be totally deleterious to the rest of the world. In fact, it would help free oil the U.S. now consumes and might even permit the nation to export more of its coal resources.

However, the virtually certain consequence of adopting such a policy would be to encourage nations desiring to expand their nuclear-power capability—and remember it is cheaper in the long run than coal—to seek even more technical-industrial help from Russia,[4] Western Europe, or Canada.

It seems unlikely, therefore, that the U.S. can do anything substantial to stem the development of worldwide nuclear power. While there are dangers in such a future world, there may be compensations, too. With ample nuclear power, nations (including the U.S.) will be less likely to consider extreme, even military, means of securing oil. Thus, in this respect nuclear power can provide a stabilizing or peaceful influence.

REFERENCES

1. S. K. Ghaswala, Science *186*, 728 (Nov. 29, 1974).
2. R. Gillette, Science *184*, 1053 (June 7, 1974).
3. Seattle Times, Dec. 17, 1974 and Seattle Post-Intelligencer, Dec. 18, 1974 (A. P.).
4. Seattle Times, Dec. 12, 1974 (UPI).

MORATORIUM MADNESS
Chapter Seventeen

Despite the defeat of the California anti-nuclear initiative in June, 1976, there is a continuing movement in the United States for moratoria on the construction of nuclear power plants. Attempts to persuade state legislatures to pass moratoria bills have so far failed, but the issue is before the public in a series of initiatives to be decided on the election ballot.

We view these moratoria proposals as sheer madness. A disturbing prospect is that the general public, with little time or background to examine the issues fully, may succumb to the irresponsible statements often made about nuclear power.

As an abstract intellectual proposition, it is painful to contemplate the blindness involved in singling out nuclear power, which is setting exemplary new standards for safety and environmental cleanliness, for the special attack. Slogans like "Split wood, not atoms" show how far the argument has strayed from any sensible regard for conservation or the environment. In recent months hundreds of miners have died in a coal mining disaster in India (the average U.S. rate is over 100 coal mining deaths per year), about twenty people died in a gas explosion in a Nebraska hotel, and 26 died in Kentucky coal mine accidents. Yet it is nuclear power alone which is claimed to be so unsafe as to require a moratorium. The world is approaching the exhaustion of its oil, yet it is only uranium which will be conserved if moratoria prevail.

But however much one may deplore the folly of the moratorium movement on an intellectual level, this is not just an abstract intellectual proposition. There are very real risks and possible losses, and these must be examined.

In this chapter, we will explore the possible consequences of a nuclear moratorium, assuming the movement spreads nationwide. Perhaps it will be easiest to analyze the consequences by categorizing three hypothetical paths which the country might take if the nuclear moratoria are adopted. We will term these: (1) the practical path, (2) the romantic path, and (3) the probable path.

The realists can take a hard look at the situation and note that there is a practical alternative to nuclear power: coal. One can debate the precise extent of our coal reserves and the relative economic costs of nuclear power and coal. Nevertheless, it is probably true that a massive commitment to coal, carried out with vigor and without great regard either to economy or to the environment, could probably solve the U.S. energy problems for at least a century.

There are some objections. Strip mining disfigures the earth and deep mining is dangerous; both problems will increase if coal expansion is very rapid. Transportation of the coal would require a major increase in rail facilities or use of our no longer plentiful water in shipping a coal slurry. Even with scrubbers to remove the sulphur dioxide, there would be a substantial death toll from coal pollution—considerably more than from nuclear-power plant emissions.

There are other objections as well, perhaps less obvious. The U.S. is unusually fortunate in its coal reserves. The rest of the world is not. Western Europe, Japan and most of the less developed world cannot take this route. We are the holders of a disproportionate share of the world's coal supply, just as the Arabs are the holders of much of the world's oil. There is little we can do about the oil, but we can try to conserve the coal for the day when it may be sorely needed by us and others as a source of oil (by liquification) or petrochemicals. It is wanton to dissipate it in the generation of electricity, as long as there are other choices.

Finally, the massive use of coal involves an environmental gamble which is poorly understood. An almost unavoidable consequence of fossil fuel burning is the emission of carbon dioxide. Even now the carbon dioxide content of the atmosphere is rising.[1] What will happen if the world continues to burn oil and coal at a rapid and probably increasing rate? We know of no atmospheric science experts who are certain whether increasing levels of carbon dioxide will be harmless, beneficial, or disastrous. One can find arguments in support of each conjecture, but if a mistake is made the recovery time of the atmosphere, if it recovers at all, will be very long. It is a dangerous chance to take in the name of safety.

The very fact that the U.S. is the holder of so much coal would, if we choose the coal route, open us to a new range of international criticisms. Already we are criticized for our rapid consumption of the world's resources. How much more ill will would we garner if, by burning ever more coal, we may damage the world's atmosphere, when we had a chance to take the clean nuclear route?

It is most unlikely, however, that the full "practical" scenario will be fulfilled. There may be some expansion of coal production, but the same environmental and economic forces which have been slowing nuclear growth will almost certainly slow coal development as well. A society which abandons nuclear energy is not likely to embrace coal with enthusiasm.

The path we call the "romantic" is the one favored by most of the vocal nuclear opponents. They have images other than coal in mind. First, they believe in conservation. Second, they believe in alternative energy sources, most commonly solar energy, but often others such as fusion, geothermal energy, and indirect solar energy in the form of wind or forest products.

It is regrettable that the vigor and dedication of the energy conservationists is not directed more toward actual energy conservation and less toward attacking energy production. There is certainly a need for using energy less wastefully. But at present, our society is better attuned to thwarting projects than to implementing them; and stopping nuclear power is easier politically to accomplish than, say, obtaining a mass transit system. It is particularly ironic that the product which will be conserved if a moratorium wins is uranium—which has no other use except for bombs.

What can we say about the hopes for the alternative energy sources? It is our belief that they are doomed to disappointment, at least if one is thinking about the problems of the next twenty-five years. It is virtually certain that there will be no fusion power before the turn of the century, if then, and it is most unlikely that in this time solar energy will be able to do more than provide a fraction of the heat needed in some of our homes. The remaining sources have prospects too small in scope to be relevant to a solution of the national energy problem. These are not our own assessments alone. They are implicit in almost all major energy studies and are the premises on which all the industrialized countries are making their energy plans.

If we abandon nuclear energy, if we do not greatly increase coal production, and if we fail in our dreams of large scale conservation or major alternative energy sources, the net result is obvious. We will import more oil. This is no fanciful speculation. At the time of the oil embargo of 1973, the country resolved to decrease its dependence on foreign oil. But after all the rhetoric, abortive plans, and high hopes, the hard reality is that we imported about as much oil in 1975 as in 1973. Alaskan oil may relieve the situation briefly, but a long-term increase is unavoidable unless effective steps are taken to change the trend.

Here is where the silent beneficiaries in the anti-nuclear movement come into their own. To replace a large nuclear power plant by an oil fired electrical generating plant requires about 10 million barrels of oil per year. Even with the price of oil as low as $10 per barrel, a year's delay of just one reactor would be worth about $100 million to whomever sells us the oil (the oil production costs being minor). It is remarkable that this economic incentive has, to date, played no apparent part in the anti-nuclear movement. At least from our own personal impressions, the one thing which would most quickly cause most anti-nuclear people to switch sides would be an offer of help from the oil interests. And yet, for all the idealism of the movement, for all its opposition to the waste of resources and (often) to big business, the greatest beneficiaries of a successful moratorium movement will be the oil producers and sellers.

Of course, such a situation would not be stable. In not very many decades the oil would just run out. Before then the country would very possibly find the burden of oil imports intolerable and, in another swing of public mood, would demand a Project Independence on a crash program basis, including a rapid return to nuclear power.

The anti-nuclear people might argue that this is fine; the delay would permit the development of "safe" nuclear power. We doubt very much that this is a valid assessment. Nuclear power is already safe. During a moratorium the government might announce specific waste disposal plans. But already there are safe methods, and the announcement of one or another specific choice will not make waste disposal any safer. Other cosmetic changes might be made to further public acceptance, as, for example, changes in insurance laws.

Under the surface, however, there would be a real loss. Trained engineers and technicians might have to leave or want to leave the nuclear field, and training of new personnel would suffer. It used to be said that a major bottleneck in our ability to build more nuclear plants was a scarcity of highly specialized welders. It is fanciful to imagine that during the period of an emotion-inspired and unpredictable moratorium there would be much push to recruit the needed new welders, or anyone else, into nuclear training programs—much less any success in getting the best people. On the contrary, the momentum of an orderly expansion would be broken. In the later transition from bust to boom, one would be starting from a weakened position, in a climate with less patience for the care and precautions now taken.

Although it is impossible to predict the economic and political developments which might accompany our greater dependence on imported oil, there is one risk which cannot be ignored. Suppose that in the future we become even more dependent on foreign oil than we presently are, say oil from Venezuela, Iran, or Indonesia. Suppose further that we believe that the oil supply is being politically threatened? Would we then stand by passively, or feel impelled to become actively engaged against the perceived internal or external threat? Similarly, would we treat the next oil embargo, if there is one, as a nuisance or as an act of war?

Our response to the oil embargo of 1973 was not encouraging. It was of short duration, at a time when our dependence on oil from the Middle East was small. Nevertheless, the inconveniences were accepted with a good deal of public ill grace and anger, including a number of reported shootings at filling stations. With the accompanying drastic increase in oil prices, the U.S. and Western Europe went into a prolonged economic recession. After the embargo there were serious suggestions that we should use military force to guarantee our access to Middle Eastern oil. If there is another oil embargo, or precipitate increase in prices, our vulnerability will be greater, and our response perhaps more immoderate.

It is perhaps too easy to invoke horror scenarios, suggesting imagined disasters in the uncertain future. However, if we were to restrict ourselves to what *probably* will happen, there would be no nuclear debate at all. Almost all the opponents concede that *probably* there will be no reactor disaster—but they say we just cannot take the risk.

The risk the world can least afford to take is that of nuclear war. If by giving up nuclear energy ourselves, we might lessen the chance of war, the sacrifice might be worthwhile. However, as we have argued in Chapter 16, the world is too much in need of nuclear energy and too committed to its use for us to be able to, or to want to, stop its spread. The best we can do is try to limit the proliferation of nuclear weapons, although it is difficult to be certain how successful we will be.

In the ever more overarmed world, there is some hope of restraint and prudence if national frustrations do not get to be too great. In a situation where a country as heavily armed as the U.S. feels threatened in its vital interests, all sorts of gambles, ordinarily unthinkable, may become plausible.

We *do not* say that a nuclear moratorium will necessarily bring on a depression or that it will precipitate World War III or another Vietnam. We *are* saying that an energy scarcity would create domestic and international tensions with unpredictable consequences. As the world advances in its ability to destroy itself, rational policy demands that we look for ways to decrease such tensions. To increase them artificially and capriciously would be tragic.

REFERENCES

1. P. V. Hobbs, H. Harrison and E. Robinson, *Atmospheric Effects of Pollutants,* Science *183*, 909 (March 8, 1974).

THE NEED FOR NUCLEAR ENERGY
Chapter Eighteen

For reasons developed in the previous chapters, we have concluded that responsible development of nuclear power is urgently needed. Our chief arguments leading to this contention should be reviewed:

1. The U.S. needs energy, and so does the rest of the world. Conservation measures may change the rate of increase of energy consumption and they may even reduce the per-capita consumption, but they cannot materially alter the basic need.

2. Fossil fuels, the present chief sources of energy, are limited. When measured on a time scale projected into the future and contrasted with man's recorded history—or even one-tenth of that time—fossil fuels cannot be considered as a solution to the world energy problem.

3. While worldwide fossil-fuel reserves, taken together, are adequate for the short run, their utilization involves special problems. The concentration of oil resources in a few countries has already created severe economic strains in the world and it has raised the prospect of even greater economic and political strains in the future. A rapidly increased rate of coal utilization carries with it substantial environmental hazards, unless adequate and expensive countermeasures are taken.

4. When we apply the two criteria of technological and economic feasibility to possible new energy sources, such as geothermal, solar, wind, and nuclear fusion, each is eliminated as a viable contender because its practicality as a large scale contributor remains to be proved. In the face of the urgency of the world's energy problems, the U.S. cannot rely on *hopes* for technological breakthroughs. Nevertheless, we should continue research and development in these areas and remain always ready to exploit any technological or economic breakthroughs which may occur.

5. Resources of uranium and thorium, together with the technically proven ability of breeder reactors to convert U-238 into usable fuel, are sufficient to satisfy the energy needs of the world for at least 100,000 years.

6. Nuclear power today is safe, ecologically acceptable, environmentally clean, economically practical, and available on a wide scale.

If the validity of each of the six arguments is established, our conclusion is inescapable, but the points supporting each argument remain the subject of debate. Each is briefly re-examined.

1. Many serious studies[1] have argued in favor of strong conservation measures. It is pointed out that Americans are extremely wasteful of energy: Our houses are too warm in the cold months and too cool in the warm months, and they are poorly insulated. Our automobiles are too numerous, they use too much gasoline, and we drive them too much; our dietary habits are too energy-expensive; our public buildings are, like our homes, too warm in winter and too cool in the summer; industry is wasteful of energy; we should shift our transportation systems to mass transit, and so on.

Each statement concerning American excesses is probably true in varying degrees, and no one can argue that considerable waste of energy is the rule in the U.S. and should be eliminated. In fact, conservation measures represent the quickest means of alleviating some of the problems. However, as a solution to the energy dilemma, conservation by itself cannot suffice, even for a relatively short time frame. And in the long run substantial reductions in energy consumption would not appreciably lengthen man's tenure on earth without new energy sources.

Very real economic barriers deter even comparatively prosperous individuals from taking obvious energy-saving steps. In order to insulate properly a home already built, to shift from an electrical-resistive heating system to a heat-pump system, or to install solar heating panels on the roof, a home owner must make a substantial capital investment, ranging from large to prohibitive. Thus, the transition to a more prudent use of energy will, at best, be gradual, relying more on better future design than on changes in existing facilities, even though such changes are immediately feasible.

In a similar vein it would be almost inconceivable to expect the owner of a 1972-model gas-eater with automatic transmission to drive it over a cliff and replace it with a $5,000, low-powered fuel-injected, 30-mile-a-gallon, miniature. That would be too much to demand even from the staunchest of conservationists. By the way, remember it does no good over-all to put the 1972 gas-eater on the used-car market; the cliff is the only way, and it has its own inherent wastefulness.

2. The precise extent of our supply of fossil fuels is certainly a matter of debate, but the recent drastic downward revision of the estimated oil reserves by the U.S. Geological Survey comes as a striking confirmation of the pessimistic estimates (see Chapter 3). Even if there were more oil and coal than indicated by the most authoritative estimates, the oil and coal should be saved for specialized purposes, rather than be consumed promiscuously in the generation of electricity.

Although coal is a large and technically feasible source of energy, it is ironic that we may reach a limit in the supply of water resources required for coal-mining and that limitation could thwart efforts to utilize coal in an ecologically acceptable and economically practical manner.[2]

Furthermore, the very fact that the U.S. is apparently blessed with up to one-third of the world's coal presents us with a serious moral question. With the real concern expressed over world-wide air pollution, including the uncertain effects of atmospheric carbon dioxide which is rising,[9] should the U.S. be contributing to this risk when we are among those who have an alternative? Are we not already sufficiently branded as the unscrupulous users of more than our share of the world's resources?

3. Since oil prices increased sharply in the fall of 1973, the U.S., Europe, and Japan have incurred oil costs in the many tens of billions of dollars. The higher prices have already created profound economic imbalance and undoubtedly have been a large contributor to the recent economic crises in both the highly industrialized and lesser-developed nations.

It is not clear what a continuation of the rising-price pattern will do to U.S. institutions and those of the still more oil-dependent European nations. But it is painfully clear that the longer the oil dependence persists, the greater will be the chance of economic chaos or military adventures. The potential consequences dwarf in magnitude those of any conceivable nuclear reactor catastrophe. The chances of horror scenarios coming true, even including nuclear war, seem much greater in the unpredictable geopolitical realm than in the much studied nuclear-reactor realm.

4. Some of the liveliest arguments in the energy controversy center around the feasibility of the alternatives. Certain groups believe, for example, that with sufficient money "clean" fusion power could be developed in a few years. However, many of these proponents of fusion seem to be unaware that one of the most promising technological possibilities appears to be hybrid-type fusion-fission reactors, which would breed both plutonium and tritium simultaneously.[3]

Even for pure deuterium fusion, high-energy neutrons are emitted which not only cause damage to materials but create radioactivity. Thus, it should be emphasized that complete "cleanliness" does not appear to be an attribute of fusion reactors.[4] On technical and economic grounds the situation also looks dim and a fusion solution appears to be far off, if ever attainable. But the authors of this book, like many others, have not given up hope for exciting breakthroughs in fusion research.

With reference to other new sources, it appears the most hopeful alternative is solar power and the authors of this book agree that all promising leads should be pursued. But again the word *hope* appears. We must explore all alternatives, but we must not base hard decisions on hope alone.

Geothermal energy will probably make a modest contribution, but its applicability may not be worldwide, and the associated environmental damage may be very serious in some regions.[5] Moreover, geothermal energy can be a serious source of air pollution.[6] In certain areas, wind power may make a small contribution.

5. There is enough uranium to sustain a large-scale nuclear program, with reactors of present design, for at least 50 years. Thorium can also help. But for the longer term there is the ultimate necessity of utilizing the breeder-type reactor. Already, strong objections have been heard to the present U.S.

breeder program, which envisages an operating demonstration breeder of about 350 megawatts.

As detailed in Chapter 14, the French breeder, called Phenix, was placed on the electrical grid in France recently. Its successful rise to nearly full power can be followed in the listings of worldwide reactor operations as published each month.[7] Furthermore, indications are that it is successfully breeding more Pu-239 than the U-235 it consumes, which breeders are supposed to do and which is the reason they are deemed crucial to the future.[8] Lastly, the reported capital investment costs of the Phenix are not much greater than for light water reactors of similar size.

6. The sixth point, concerning the safe and benign character of nuclear energy, is obviously the most controversial. Nuclear power could certainly have been dangerous, had it been introduced carelessly. We believe, however, that the safety measures which have been imposed have not merely made it safe and clean; they have made it extraordinarily safe and clean. The widespread doubts which exist on this score seem to us to have little reasonable basis. Instead the country appears to be slipping into a mood of nuclear fear which is out of touch with reality. Not only has a sense of perspective been lost; it seems almost ignoble to try to restore it.

What people worry about is often a matter of fashion. One can look for communists under every bed, or CIA agents. We can build bomb shelters in panic, or trust detente. We can shrug off 50,000 virtually certain automobile deaths, or become paralyzed by the prospect of 5,000 almost impossible nuclear deaths. We now receive less than one thousandth as much radiation from the nuclear power program as from medical x-rays. If the nuclear program is in full swing in the year 2000, it will give us less than one-hundredth as much radiation as will our medical programs. Yet nuclear power is a matter of national alarm, and medical radiation exposure is gladly accepted by the public.

These national trends are not determined by the objective situation. They are established by some subtle blend of actual circumstances, social psychology, and the swing of the pendulum. Once the mood is set, it is awkward to be too far out of step.

The good grey New York Times, for example, cautiously describes licensed nuclear reactors in the United States as follows: "Although some accidents have occurred, none has been catastrophic." (New York Times, The Week in Review, February 15, 1976.) This phrasing is totally misleading. It conjures up the thought that we have not yet wiped out, say, Pennsylvania. But in reality we have done far better than that. Not only have *catastrophes* been avoided; there has never been a *single* person killed or injured by radiation from any of these plants. If it is human life we are concerned with, the sentence should have read: "Although some accidents have occurred, none has hurt anyone." Less fashionable phraseology, but more precise.

If we turn to the actual performance of nuclear reactors we find that the environmental impacts *are* small and radioactive emissions *are* negligible. Although the plants are more complicated than fossil-fuel plants, nuclear-reactor technology is already available and proved efficient. The waste material is very small in volume, and we *do* know how to take care of it and store it. Furthermore, the experimental test of a long-term storage method—in case all else is deemed unacceptable—has been made for us by the ancient Egyptians: Their elegant pyramids are still standing for all to see as convincing evidence of the durability of man-made structures. If the interiors of similar structures were filled with old radioactive debris, one could not care less and could climb their sides without worry, because comparatively little shielding would be required against the residual debris.

Could it be that one of the most important legacies left to us by the great Egyptian civilization is the example of durability provided by the pyramids? It is an intriguing thought, but it is likely that we will instead follow the reasonable course of using deep salt deposits, which geologists tell us are even more durable than stone or concrete pyramids.

Among the controversial questions swirling around the subject of nuclear power, the one that is uppermost in many people's minds is reactor safety. In our view, the Reactor Safety Study,[10] or Rasmussen Report, has put the worry to rest. It is a long and technical document, over 5 inches thick, and it presents a formidable task for the most expert of experts to digest its details. Its summary, however, is clearly written, and many non-experts may find it informative and reassuring.

Unfortunately, few persons have time to pursue this sort of study or even much more modest studies. Instead, the issue is "settled" in the public mind by headlines, or, at best, brief news items and fragments of debate. In such a climate, nuclear power faces the possibility of public rejection—especially if more news value is given a minor failure in a nuclear plant that injures no one than to, say, a natural-gas explosion in which people are killed or seriously injured, or if more attention is paid to one anti-nuclear engineer than to a hundred pro-nuclear engineers. Nuclear power will face certain rejection if society adopts tacit ground rules under which absolute perfection and unanimity are demanded for nuclear power, while the consequences of *not* deploying nuclear power are ignored.

In the course of our study, we have questioned particularly why environmental and conservation groups have come to oppose nuclear power. Both of us have always been conservationists, and one author is a past member of two of the largest mountaineering organizations in the nation, is still a member of a third, and is a contributor to and a member of several conservation societies.

We sympathize with the genuine concerns of the conservationists, but we believe firmly that when all the facts and alternatives are carefully examined, these concerns will disappear. In our view, the goals and desires of environmentalists and conservationists are far better served by nuclear energy than by any other approach that is technologically and economically feasible today. Energy conservation by itself simply cannot be a solution.

It is also asserted on occasion that nuclear power cannot make a significant contribution to meeting the great need for energy in the short-term period— say the next decade. It not only can do so; it already has proved itself to be a significant energy source. And with a national will to accelerate nuclear-plant construction, the lead time or over-all time from conception to operation can be reduced from the present ten years to about six.

In addition to the opposition from various protest groups, another impediment has arisen to the immediate acceleration of nuclear power—the lack of capital for investment. In sharp contrast to the supremely well-financed efforts to expand oil production through offshore drilling, construction of the Alaska pipeline, and many other means, the electric utilities are unable to attract sufficient credit or to utilize their non-existent excess profits for expansion, as the oil companies can. It is this squeeze that has been a major cause in recent years for the setbacks in the planning and construction of many new nuclear and, though less well known, coal-fired plants.[11]

On a national scale, this predicament leads to a serious misuse of economic resources. American payments to the oil-producing countries—about $25 billion to $30 billion a year and climbing[12]—are equal to the capital costs for about 30 new power plants each year. Thirty power plants, if oil-fired, would use about a million barrels of oil a day, or about one-sixth the present import rate. Therefore, it is clear that investment in nuclear-power plants at this time could mean enormous oil savings in the future, with a simultaneously dramatic improvement in the nation's international economic position.

Much of the opposition to nuclear power grew out of criticism of the Atomic Energy Commission. It is readily acknowledged that mistakes were made. No new technology, in America's spectacular technological history, has been free of mistakes. Mistakes even provide an important learning technique. But the mistakes have led to none of the disasters which critics keep invoking, and overall, the AEC and the nuclear industry achieved a safety record which is unparalleled.

On the local and state level, too, a climate of distrust and misunderstanding was generated by friction between concerned citizens and insensitive nuclear industry and electric utility officials, often aggravated by (for example) poor choices for power plant siting.

With the phasing out of the AEC, its replacement by the Energy Research and Development Administration (ERDA) and the Nuclear Regulatory Commission (NRC), and with greater efforts for mutual understanding at all levels, the nation can hopefully look forward to a new era in which past institutional antagonisms are forgotten.

In final recapitulation, we believe the greatest over-all safety for the U.S. now and for future generations lies in building a stable, prosperous, and peaceful society in America and the rest of the world. We do not see how such a society can be created if the U.S. does not maintain a continuing and, in fact, increasing, supply of energy. It is needed not merely to maintain the pattern of life of middle-class America, but to meet the requirements of her poor, to help relieve the extremes of poverty elsewhere, and to make allowances for the inevitable increase in world population, at least for the short run. At present, nuclear power offers the surest, safest means of providing much of the required energy and of helping reach these goals.

REFERENCES

1. For example: Ford Foundation, Energy Policy Project, *A Time to Choose: America's Energy Future,* Ballinger Pub. Co., Cambridge (1974).

2. R. Gillette, Science *181*, 525 (Aug. 10, 1973), a review of a National Academy Report.

3. B. R. Leonard, Jr., *A Review of Fusion-Fission (Hybrid) Concepts,* Nuclear Technology *20,* 161 (1973).

4. B. L. Cohen, Physics Today, *27*, 15 (Nov. 1974).

5. See, for example, editorial in Seattle Times, Feb. 2, 1974.

6. R. C. Axtman, *Environmental Impact of a Geothermal Power Plant,* Science *187*, 795 (Mar. 1975).

7. These operations are reported in Nucleonics Week.

8. William D. Metz, *European Breeders (I): France Leads the Way,* Science *190,* 1279 (Dec. 26, 1975).

9. P. V. Hobbs, H. Harrison, and E. Robinson, *Atmospheric Effects of Pollutants,* Science *183,* 909 (Mar. 8, 1974).

10. Ref. 5, Chapter 13.

11. Federal Energy Administration, Region 10, press release July 11, 1975.

12. Frank Zarb (Federal Energy Administrator), Seattle Times, June 8, 1976 (UPI).

ENERGY CONVERSION CHART††
Appendix A

The chart is similar to a road mileage chart. However, for distances it doesn't matter whether you go from New York to Washington or Washington to New York. Thus, a mileage chart has numbers in only one half of the intersections of the rows and columns. But a chart for converting from one measure to another needs numbers in both "directions." To illustrate: One yard equals three feet, whereas one foot equals one-third of a yard.

The chart is arranged so that one unit of a quantity of energy in a horizontal row equals that number of units of energy in a vertical column shown in the intersecting square.

To save writing zeros, the numbers are shown in scientific notation. They are to be multiplied (+) or divided (−) by the power of 10 indicated by the lower number in the square. Thus, $10°=1$; $10^1=10$; $10^{-2}=1/100$, and so on.

††Based on a more extensive chart prepared by Kelly C. Green.

Example:

One gallon of gasoline equals 5.78/1000 tons of coal (5.78×10^{-3}).

	foot-pounds	kilowatt-hours	Calories	tons of coal	barrels of oil	gallons of gasoline	kilograms of uranium	B.T.U.'s	cubic feet Natural Gas
1 foot-pound	1	3.77 −7	3.24 −4	4.98 −11	2.22 −10	8.62 −9	2.03 −14	1.29 −3	1.22 −6
1 kilowatt-hour***	2.66 +6	1	8.60 +2	1.32 −4	5.88 −4	2.29 −2	5.38 −8	3.41 +3	3.25 0
1 kilocalorie or Calorie	3.09 +3	1.16 −3	1	1.54 −7	6.84 −7	2.66 −5	6.26 −11	3.97 0	3.78 −3
1 ton of coal**	2.01 +10	7.56 +3	6.50 +6	1	4.45 0	1.73 +2	1.07 −4	2.58 +7	2.46 +4
1 barrel of crude oil†	4.51 +9	1.70 +3	1.46 +6	2.25 −1	1	3.89 +1	9.15 −5	5.80 +6	5.52 +3
1 gallon of gasoline	1.16 +8	4.37 +1	3.76 +4	5.78 −3	2.57 −2	1	2.35 −6	1.49 +5	1.42 +2
1 kilogram of uranium*	4.92 +13	1.86 +7	1.60 +10	2.46 +3	1.09 +4	4.26 +5	1	6.34 +10	6.04 +7
1 British Thermal Unit	7.78 +2	2.93 −4	2.52 −1	3.85 −8	1.72 −7	6.71 −6	1.58 −11	1	9.25 −4
1 cubic foot Natural Gas	8.17 +5	3.08 −1	2.65 +1	4.07 −5	1.81 −4	7.04 −3	1.66 −8	1.05 +3	1

*Total energy content of U via breeders. Pu-239 and U-233 are very nearly equal to uranium.
**2000 pounds.
***kilo = 1000; mega = 1,000,000.
†One barrel is approximately 42 gallons, U.S. measure.
Conversion factors for coal, oil, and gasoline are for average types. All conversion factors assume direct conversion from one unit to another. No account is taken of, for example, thermal to electric conversion efficiency.

CALCULATION OF INCREASE OF TEMPERATURE OF THE EARTH
Appendix B

The rate of energy consumption in the U.S. is about 10 kilowatts/person: 10^4 watts per person.

Present world population: 4×10^9 (4 billion)

Therefore total power for triple present population, all at U.S. rate: $3 \times 4 \times 10^9 \times 10^4$ watts $= 1.2 \times 10^{14}$ watts.

Approximate power received from sun: 1.3×10^{17} watts.

But power radiated by earth is approximately $P = \sigma T^4$, applying the law of radiating bodies. (σ is a constant.)

By differentiating P and dividing by P it follows that

$$\Delta P/P = 4 \, \Delta T/T = \frac{1.2 \times 10^{14}}{1.3 \times 10^{17}} \simeq 10^{-3}$$

and therefore $\Delta T = 10^{-3} \times T/4 = 10^{-3} \times (\frac{1}{4}) \times 300 \simeq 0.07°$ Celsius (or about $0.13°$ Fahrenheit).

The 300° in the calculation is the average temperature of the earth in absolute degrees; i.e., 300° above the absolute zero of temperature.

The result, 0.13° Fahrenheit, shows that there is little likelihood that the earth would heat up appreciably even if everybody used energy at the U.S. rate, and if the population of the world were tripled.

THE LINEARITY ASSUMPTION FOR RADIATION EFFECTS
Appendix C

Most calculations of the probable consequences resulting from the release of radioactivity are based upon the *linearity assumption* or *hypothesis*.[1] Reference was made to it in Chapter 13 on reactor safety; it is also applicable to the issue of radiation hazards in normal operation (Chapter 10) and in fact to virtually all situations where problems of radiation hazard arise.

It is evident, then, that the linearity assumption is a basic ingredient in consideration of many controversial aspects of nuclear power. Therefore, it is important to understand what it means, the experimental base upon which it rests, and *why it is an assumption.*

It is known that if people or animals receive a sufficiently large radiation dose within a short period of time, they will die. For humans, there is a high chance of death above 400,000 mrem. (The unit used here, the millirem or mrem, has been introduced in Chapter 10.) For lower doses, down to about 150,000 mrem, exposed individuals have a *chance* of dying; the percentage who will die depends upon the dose. The data for people is, fortunately, not very extensive, but it is adequate to establish these numbers as approximate guides.

For still smaller doses, there are no deaths from radiation sickness, but there are delayed effects of cancer production and, probably, of genetic damage. The cancers typically appear after one or more decades; the genetic effects appear in the next generation, and, at a decreasing rate, in subsequent generations.

The evidence for radiation induced cancer comes mostly from animal experiments, from the experience of humans who received high doses in medical treatments, and from the atomic bomb victims. The evidence for genetic effects comes *only* from animal experiments, because to date there is no evidence of observable genetic radiation damage in humans.

In all these cases, the conclusions are based on experience at relatively high radiation levels. For example, the known cancers in humans were observed in situations where the exposures were typically above 50,000 mrem and were delivered at rates of over 1,000 mrem per minute.[2] For this exposure there is about a ½ to 1 percent chance of cancer production. A person living right next to a nuclear power plant, on the other hand, will receive not more than about 10 mrem, delivered during a year. Similarly, people exposed to radioactive wastes and most people exposed in case of a reactor accident will receive doses far below 50,000 mrem. How can effects of much lower doses be estimated?

This is where the linearity assumption comes in. According to this assumption, it does not matter if 1 person receives a given dose or 10,000 people receive a dose 10,000 times smaller. There is the same total chance of cancer. Thus, using the numbers in the preceding paragraph, one would infer that if 10,000 people each receive a dose of 10 mrem, there is a 1 to 2 percent chance that *one* of them will get cancer from the radiation. In this assumption, the dose rate does not matter. It is just as bad to get the radiation delivered slowly over a year as to get it all at once. This part of the linearity assumption might be called the time-exposure reciprocity rule—a rule, familiar to many amateur photographers, applying to exposure of photographic film.

Also implicit in this estimate is a second, closely related, assumption—namely that no threshold exists. The irradiated person or animal is assumed to possess no tolerance level. Thus, it is assumed that as far as radiation induced cancer is concerned the body has no recuperative powers.

Exposure to other bodily abuses does not exhibit the same linearity effect. If one drinks a fifth of whiskey in a single draught on an empty stomach, one is running a substantial risk of death. However, the same quantity of whiskey, consumed over a period of time, has a minor effect, and for some a pleasant one. Furthermore, we do not find any obvious evidence of cumulative detrimental effects from the total amount of whiskey consumed, regardless of the size of the separate doses unless total consumption is excessive. In short, the body appears to recover from the deleterious effects of whiskey. Of course, many other examples could be cited, such as eating too much salt too fast as opposed to eating salt at a reasonable rate.

These are examples of effects exhibiting thresholds. The fact that people *do* die in a short time, if irradiated heavily, but do not, if irradiated lightly, is itself a threshold effect. Other such thresholds exist for radiation, such as reddening of the skin and vomiting.

Now, if there is no firm experimental basis for the linearity assumption at low dose rates and for the assumption that no threshold exists for radiation doses, why are they adopted as a basis for calculations—such as the one in the American Physical Society Reactor Safety Study leading to such predictions as 10,000 deaths (see Chapter 13)?

The most authoritative endorsement of the linearity assumption, or hypothesis, comes from the 1972 Report of the Advisory Committee on the Biological Effects of Ionizing Radiation of the National Academy of Sciences and the National Research Council (the BEIR Report, reference 1). The BEIR Report discusses arguments for and against the linearity hypothesis, stressing the uncertainties. In a discussion of the relationship between radiation dose and cancer production, it states (page 97):

> In view of the gaps in our understanding of radiation carcinogenesis in man, and in view of its more conservative implications, the linear, nonthreshold hypothesis warrants use in determining public policy on radiation protection; however, explicit explanation and qualification of the assumptions and procedures involved in such risk estimates are called for to prevent their acceptance as scientific dogma.

Other authorities state matters with a different emphasis. Thus, Laurison S. Taylor, President of the National Council on Radiation Protection and Measurements (NCRP) explains the linearity assumption as[3]

> ... made with the maximum credible conservatism so that any calculations using them would result in grossly overestimating or maximizing any possible effects.

This view is reaffirmed in a 1975 Report of the NCRP.[4] Introducing a critical analysis of the existing data and of various studies, including the BEIR Report, it states (page 2):

> The NCRP continues to hold the view that risk estimates for radiogenic cancers at low doses and low dose rates derived on the basis of linear (proportional) extrapolation from the rising portions of the dose-incidence curve at high doses and high dose rates, as described and discussed in subsequent sections of this report, cannot be expected to provide realistic estimates of the actual risks from low level, low-LET [low linear energy transfer] radiations, and have such a high probability of overestimating the actual risk as to be of only marginal value, if any, for purposes of *realistic* risk-benefit evaluation.

Doubts about the validity of the linearity assumption have also been recently expressed in a report of the British Medical Research Council.[5]

One can speculate about the reasons for the differences in tone and emphasis between the BEIR Report and the NCRP Report. Both were prepared by prestigious groups, with a large overlap in personnel. (About half the members of the Advisory Committee preparing the BEIR Report were members of the NCRP at the time the NCRP Report was issued.) The NCRP Report came out about two years after the BEIR Report and cites some newer data, but the overall body of information worked with was largely the same.

It is conceivable that the greatest difference was in the "moral" the authors intended to convey. The BEIR Report, issued at a time when there was some feeling that federal agencies were unduly complacent about low radiation doses received by the general population, states (page 2):

> The present guides of 170 mrem/yr grew out of an effort to balance societal needs against genetic risks. It appears that these needs can be met with far lower average exposures and lower genetic and somatic risk than permitted by the current Radiation Protection Guide. To this extent, the Guide is unnecessarily high.

The NCRP Report, issued at a slightly later time when there was considerable public dispute about hazards from nuclear power at radiation levels well below 170 mrem/yr, states (page 4):

> The NCRP wishes to caution governmental policy-making agencies of the unreasonableness of interpreting or assuming "upper limit" estimates of carcinogenic risks at low radiation levels, derived by linear extrapolation from data obtained at high doses and dose rates, as actual risks, and of basing unduly restrictive policies on such an interpretation or assumption. The NCRP has always endeavored to insure public awareness of the hazards of ionizing radiation, but it has been equally determined to insure that such hazards are not greatly overestimated. Undue concern, as well as carelessness with regard to radiation hazards, is considered detrimental to the public interest.

Perhaps the easiest conclusion to draw from both the BEIR Report and the NCRP Report is that there is a great deal of uncertainty. On the one hand, the linearity assumption gives a plausible estimate of the *maximum* number of cancers or genetic effects which may be produced by low-level radiation. On the other hand, it may grossly overestimate the effects. When the linearity assumption is used, as in the American Physical Society Study, it cannot be concluded that 10,000 people *will* die of radiation-induced cancer following a large accident. One can only conclude that they *might* die or *might* not die; no solid evidence is available to prove either statement. However, it is clear that the linearity assumption can be misunderstood when the word *might* is converted to *will* in discussing low-level radiation effects.

In view of the obvious importance of the linearity assumption, together with its sub-assumptions, it would appear to be very important to verify it experimentally. However, an illustration will show why this is very difficult.

According to the assumption, a dose of 100 millirems administered uniformly each year to 200 million people will produce about 3,500 additional cancer deaths a year.[6] That is the same as 18 deaths a year for one million people.

The difference in radiation received by the approximately one million persons living in the Denver area and the same number living in the New Orleans area is about 100 millirems per year. Thus, we should annually expect to find 18 more cancer deaths in Denver than in New Orleans—clearly a hopeless task, since almost 2,000 die of cancer each year in each area. A variation of 1 percent is statistically meaningless here even if the populations and other (non-radiation) environmental features were perfectly matched.

Thus, at this time the correctness or incorrectness of the linearity assumption cannot be unambiguously determined. It was introduced as a prudent guide for setting radiation exposure standards. It also can provide upper limits on the consequences arising from specific situations of exposure. However, it appears inappropriate to regard the linearity hypothesis as an established fact, and to accept it as giving realistic estimates of actual and inevitable consequences.

REFERENCES

1. See *The Effects on Populations of Exposure to Low Levels of Ionizing Radiation*, National Academy of Sciences-National Research Council (November, 1972). (This report is commonly referred to as the BEIR Report.)

2. Ibid, p. 86.

3. L.S. Taylor, *The Origin and Significance of Radiation Dose Limits for the Population*, National Council on Radiation Protection and Measurements. WASH-1336 (August, 1973).

4. *Review of the Current State of Radiation Protection Philosophy*, National Council on Radiation Protection and Measurements, NCRP Report No. 43 (January, 1975).

5. *The Toxicity of Plutonium*, Her Majesty's Stationary Office, London (1975), p. 7.

6. Ref. 1, page 2.

TEST OF THE EMERGENCY CORE COOLING SYSTEM
Appendix D

One of the most commonly raised issues in the debate over reactor safety concerns the reliability of the emergency core cooling system, the ECCS. As explained in Chapter 13, a full scale experimental test of the ECCS involving a guillotine-like main pipe break has never been made. Such a test would be exceedingly difficult and possibly very destructive. In this appendix we discuss some arguments for and against a full scale test, and attempt to put the debate in reasonable context.

The ECCS consists of many components. Most of the separate components can be tested individually. Many of them are duplicated so that two or more must fail simultaneously in order for the system as a whole to fail. Thus, the probability of simultaneous failure of a whole chain of components can be calculated with considerable reliability, provided each individual failure rate is known. In Chapter 13 the method was illustrated with an example from automobile experience.

Calculations made in this manner lead reactor safety engineers to be confident that for any loss of cooling caused by small or medium sized breaks in the cooling system, the ECCS indeed will work as intended. Fortunately, calculations and experience also show that the chance of small pipe breaks is much greater than the chance of very large breaks, so the importance of having an ECCS diminishes as the size of the postulated break increases.

If every operational step of the ECCS could be tested adequately, there would be essentially no uncertainty, and no debate. However, there is one important case which relies heavily on calculations and extrapolations from small scale "mock-up" tests. Some of these small scale tests have failed, and some of the calculations cast doubt on the ability of the ECCS to properly provide emergency cooling water. However, it is not clear if these failures demonstrated an inadequacy in the ECCS or whether the tests themselves were inadequately conceived.

The greatest potential trouble arises for a very large failure in one of the 3-feet-in-diameter, 3-inch-thick main cooling pipes. The water in the reactor vessel, which normally cannot boil because it is under great pressure, is then suddenly at much lower pressure. Steam develops very rapidly and blows the water out of the vessel. At a certain stage, the ECCS is designed automatically to inject new water (together with a boron solution to prevent the core from starting up again) from a large pressurized storage tank into the reactor vessel. But the injected water may be forced by the violent boiling to cross the top of the vessel and be directed uselessly into the other undamaged pipes, and thus never reach the reactor core.

As noted above, some small scale tests have failed. As a result, modifications have been made in both design of the ECCS and in the computer calculations used to predict performance. In addition, a medium scale test, in a reactor of about 50 megawatts, is planned for the near future. It is called "LOFT" for loss-of-fluid test.

With the doubts cast on the ability of the ECCS to provide cooling in the event of large pipe failures, the safety of large nuclear reactors might appear to be in serious question indeed. However, the debate must be placed in proper context by asking three questions: Has the ECCS ever been required in operating reactors? What is the probability of such a great cooling line failure? By how much are the chances of a serious accident increased if the ECCS doesn't work in this particular case?

Since the answer to the first question is no, we turn to a discussion of the second.

Experience with large high pressure pipes is very extensive. For example, coal-fired plants of similar size require similar piping. There have been *no* failures in pipes built to standards set for nuclear reactors. Based on these experiences, the Reactor Safety Study estimated the probability of a total pipe failure which could cause a large cooling loss to be one in ten thousand per reactor-year. With such a low probability of failure, coupled with less than one chance in one hundred of serious consequences even if a pipe did fail, the importance of the ECCS is greatly diminished.[1]

By how much are the chances of an accident increased if the ECCS does fail for a hypothetical large pipe failure? The Reactor Safety Study estimated this increase to be between a factor of about two and four, with a maximum of a factor of ten.[2] The effect, therefore, is to make the probability of a major nuclear accident as much as ten times more probable than a major meteor impact having similar consequences; but the nuclear accident is still 100 times less probable than a dam failure having equally severe consequences.

Although a full scale test would be difficult, it would not be impossible, and despite the reassuring numbers given above, society could decide it is still worth it. One can speculate on how such a test could be made.

Obviously, an existing reactor would not be used, because of the possibility of contaminating the surrounding area, but a full scale reactor conceivably could be built in a remote place, so that if the test resulted in release of radioactivity, damage would be minimized. The cost would not be excessive—perhaps one or two billion dollars, or about the cost of a Trident submarine or two. The motivation in each case is the same—to enhance our security. The only difference is that the Trident is built with the intent of enhancing national security by presumably creating a situation discouraging its use; whereas the test reactor would be built in hopes of enhancing security by increasing the acceptibility and use of nuclear power.

An analogy with the proposed test can be drawn. The thermonuclear or hydrogen bombs *had* to be tested to find out if they worked.[3] Due to their magnitude, the tests were carried out on remote Pacific islands. Moreover, the amount of radioactive *fission* products from uranium[4] dispersed in the atmosphere by the larger tests was roughly equal to the total amount present in a 1,000 Mw(e) reactor core. Thus, though a reactor which failed the test under the worst conditions could conceivably emit similar quantities of radioactive debris, the test would be far less violent than the bomb tests and would not disperse radioactivity as widely. Thus, it is clear that society *could* make the destructive reactor test.

If a successful test were made under controlled conditions on a test reactor, some doubt would still linger as to whether all reactors would behave similarly. Nor would the test have included all situations; for example, the operating personnel would be different.

For each of us who drives automobiles, there exists an analogous situation. Almost every driver has conceived of the possibility that he and his automobile will be called upon some day to make a "panic" stop while traveling at a speed of 60 or more miles an hour. Despite that possibility, how many of us have ever practiced making such a stop in advance so we would be ready for it in an emergency? Is the remote chance of needing to make a panic stop worth the small risk that something will go wrong when we practice it?

Those few of us who have practiced the maneuver have discovered we cannot quite bring outselves to make a legitimate panic stop. We retain a slight uncertainty over our capability, or the automobile's, to take the strain. We stop "as fast as we can," and although we nurture a belief that we could do much better if we "really had to," we are not quite positive.

Among the safety devices built into our newer automobiles is a dual brake system; therefore, were we to over-stress the hydraulic brake systems in a panic stop, perhaps one crossed pair of front-and-back brakes might fail. But then, with unbalanced steering, will the car swerve into oncoming traffic or into an obstruction? The question cannot be fully answered for each individual, and the definitive test cannot be made without incurring excessive personal risk.

Of course, automobile engineers do make such tests, presumably under controlled conditions, and these tests are analogous to our proposed billion-dollar remote-location ECCS test. But just as this test does not include the same total situation, neither does the automobile engineer's test; that is, we (the real drivers) are left out.

If we choose to accept the predicted accident probabilities for the chance the ECCS will be called upon, for the chance of its successful operation (if called upon), and the consequences in case it doesn't work, then it is clear that the reduction in loss of human lives by investing efforts toward greater automobile safety is much greater than by performing the postulated total ECCS test.

Indeed, the fact that we *can* make our own automobile braking test at *moderate* braking effort gives us confidence that we probably can perform a panic stop. Similarly, the calculations give us confidence that the ECCS *probably* will work.

To summarize, man and his technology, operating in concert, can devise safety systems which, by all reasoning, have a high probability of operating successfully in the emergencies for which they are designed. But sometimes one cannot achieve absolutely total assurance of their reliability through direct experimental testing.

For the worst possible scenarios, the ECCS probably falls in this category. It has a good chance of working, but it is not *absolutely* certain that it will. Fortunately, the increased risks, even if it does not, are indeed not very great when analyzed dispassionately.

REFERENCES

1. WASH-1400, *Reactor Safety Study*, Appendix III, p. 74ff.

2. Ibid, Appendix V, p. 34.

3. Herbert York, *The Advisors: Oppenheimer, Teller, and the Super Bomb*, W. H. Freeman (1976), p. 50 and 107.

4. Ibid, p. 90. Four hundred kilograms of fission products were produced in the "Bravo" test. This is about the same as the inventory in a 1000 MW(e) reactor core after lengthy operation. The thermonuclear bomb is really a "hybrid reactor," utilizing both fission and fusion reactions. See Ref. 3, p. 107.

NUCLEAR LIABILITY INSURANCE AND THE PRICE-ANDERSON ACT
Appendix E

Nuclear power has just the ingredients to make it appear to be a liability insurance nightmare:

- There have been no major accidents, so there is little actuarial experience to go on.
- The best predictions indicate that an accident is very unlikely, so there is no objective justification for substantial liability insurance premiums.
- If an accident should occur, the property damage could be very large, so no insurance company or group of companies could cope with the possible claims.
- In the event of an accident, the chief health consequence would be a small increase in the cancer rate for the exposed population, so it would be virtually impossible to identify the individual victims.
- In the event of an accident, many simultaneous failures would have been involved, so that the assignment of blame would be difficult.

Nevertheless, insurance is needed. We will consider some aspects of how this problem has been settled.

First, let us try to obtain an estimate of a "fair" premium, which the utilities should pay to get adequate insurance coverage. Averaging over large and small accidents, the Reactor Safety Study (the Rasmussen Report) estimates the average annual property damage from reactor accidents to be $2 million for 100 reactors.[1] This would imply insurance charges of $2 million a year for the 100 reactors, or $20,000 per year per reactor.

There is also damage to life and health to consider. In amount, this would not add greatly to the liability for property damage because the predicted number of cases is small. There is a complexity, however, because the most severe medical consequences are those of cancer induction, with the cancers showing up many years later. For every cancer caused by the nuclear accident, there will be 10 or 100 (or more, depending upon the size of the accident) which would have occurred anyway in the exposed population. Who then gets compensated?

No doubt, as society becomes more conscious of the problems of environmental pollution, there will be policies established to provide compensation for unknown "statistical" victims of all sorts of industrial pollution. But at the moment there seems to be no satisfactory solution to this broad problem, and it would have to be faced if there were a large nuclear accident.

The existing nuclear insurance system is governed by the Price-Anderson Act, first passed in 1957. At the end of 1975 Congress extended it, with amendments, to be in effect until August, 1987.[2] It will therefore be the governing law for some time, barring further changes.

In brief, Price-Anderson makes it relatively easy for claimants to receive compensation up to some specified total limit, currently $560 million for a single accident, but frees the nuclear industry and the government from liability for compensation beyond this limit.

At present the $560 million is covered by private insurance pools of $125 million and an indemnity commitment by the federal government for the remaining $435 million. An interesting provision in the new legislation will progressively reduce and eventually eliminate the federal indemnity commitment, replacing it with a growing utility industry commitment.

The way this is to be accomplished is by the introduction of a "deferred retrospective premium." The words are forbidding, but the idea is simple. Because a nuclear accident is very unlikely, there is no need to extract all premiums ahead of time. Should there be an accident, then each utility will be assessed by the amount of the retrospective premium. The amount of this premium has not yet been specified; by the end of 1976 the Nuclear Regulatory Commission is required to establish it "at some level between $2 million and $5 million."[2] Suppose $3 million is picked as the final number. Further, suppose that in 1980 there is an accident and there are then 90 operating reactors. The operators of these reactors would each be required to contribute $3 million. This would mean $270 million from this pool, reducing the federal indemnity from $435 million to $165 million. When there are 145 reactors, the federal indemnity becomes zero.

Beyond 145 reactors, the ceiling would start to rise above $560 million. Thus, for example, suppose there is an accident in 1990, when there are, say, 400 operating reactors. Then the retrospective premium pool, at $3 million per reactor, would be $1,200 million. Together with the "ordinary" private insurance of $125 million, the maximum liability would be $1,325 million, more than double the current maximum.

Of course, a ceiling of $560 million, or even $1.3 billion, is not enough to cover the worst possible accident, and above the ceiling the industry and the government have no legal liability. This fact is the basis for much of the enunciated criticism of the Price-Anderson legislation.

The specific present ceiling of $560 million appears arbitrary, and, as with any arbitrary number, it is hard to defend. Its roots are historical; in 1957, Congress established requirements for $60 million of private insurance and the round number of $500 million for federal indemnity. One might imagine that with inflation, if for no other reason, the ceiling might have been subsequently raised, but it has not been. However, the existence of a ceiling has not been viewed by Congress as a significant limitation on the public's real protection. Thus, the Joint Committee on Atomic Energy stated in 1965:[3]

> In this connection, the committee has carefully considered the subject of the limitation of liability which is contained in the Price-Anderson legislation. Under the bill recommended by the committee, this limitation would continue to be set at the total amount of financial protection required plus the governmental indemnity, but in no event to exceed $560,000,000. It is the committee's view that this limitation does not, as a practical matter, detract from the public protection afforded by this legislation. In the first place, the likelihood of an accident occurring which would result in claims exceeding the sum of the financial protection required and the governmental indemnity is exceedingly remote, albeit theoretically possible. Perhaps more important, in the event of a national disaster of this magnitude, it is obvious that Congress would have to review the problem and take appropriate action. The history of other natural or man-made disasters, such as the Texas City incident, bears this out. The limitation of liability serves primarily as a device for facilitating further congressional review of such a situation, rather than an ultimate bar to further relief of the public.

In short, if the Price-Anderson ceiling is exceeded, Congress would be expected to improvise, as it does for any other sort of national disaster.

Such a view certainly is consistent with our expectations in other areas. There was no prior legislation requiring the federal government to compensate coal miners for black lung disease, yet we are now paying about $1 billion *each year* in compensation to former miners. Natural disasters, such as hurricanes or earthquakes, or man-related disasters such as dam failures could cause far more than $1 billion in damages. It is very much more likely that such disasters will occur than that there will be a major nuclear accident.[4] Yet no insurance scheme could cope with disasters of such magnitude or pretends to. We rely on proper action after the event by a humane and reasonably prosperous society.

There are other financial disasters against which we have no insurance. When the price of imported oil rises by $10 per barrel, we lose over $20 billion per year. There is no insurance against that loss and society as a whole pays the bill. The comparison with oil has another relevance. We currently are using about 500 million barrels of oil a year for generating electricity.[5] This costs about $5 billion annually, and is equivalent to about one-fifth of our oil imports. The same amount of electricity could be generated by 50 reactors. In

3 years, these reactors would save us enough oil to "cover" the $14 billion property damage[6] of even the worst-case nuclear accident.

In considering the implications of the liability ceiling, we should also ask what the likelihood is it will be exceeded. According to the Reactor Safety Study, with 100 operating reactors there is less than one chance in 3,000 per year of property damage greater than $560 million.[4] Thus, it is most improbable that Congress will ever be called upon to face up to the insurance problems created by an accident which exceeds the Price-Anderson liability limitations.

To proceed to another aspect of the insurance controversy, it is sometimes suggested that the absence of unlimited liability will tend to make the nuclear industry more careless. We believe this worry to be groundless, for the following reasons: (1) Its liability is still $125 million, which itself is a substantial deterrent. (2) The main financial loss, in all but the very largest accidents, comes from damage to the reactor and from the loss of generating capacity. Thus, liability aside, there is an enormous financial incentive for safe, accident-free operation. In the Browns Ferry accident, for example, there was no damage outside the plant and no problem of liability. Yet the Tennessee Valley Authority is having to spend over $100 million for alternative electricity sources, while the plants are being repaired. (3) The main guarantor of reactor safety must be the Nuclear Regulatory Commission. Good safety practices must be enforced by stringent federal regulation, not by reliance on the presumed financial prudence of the industry.

Finally it is suggested that the liability limits imply that the insurance companies are skeptical about the reactor safety claims of the Rasmussen Report and of the sponsoring federal agencies. Even if one granted the unreasonable hypothesis that insurance companies have better insight into reactor safety than does the Nuclear Regulatory Commission, the limit of $125 million per reactor on liability (plus up to $175 million coverage on the reactor and other property of the utilities[7]) represents a very substantial commitment. For the over 50 reactors in operation, this means an overall liability commitment of over $6 billion. The coverage for a single accident is unusually high. Of course, it is typical, not unusual, for limits to be placed on insurance coverage.

There is one difference, however. Under the Price-Anderson Act, once the maximum insurance coverage has been paid out, the utilities and the manufacturing companies are free of further legal liability. In the absence of such a provision, although the injured parties might not be able to collect compensation from a bankrupt company, they could have the satisfaction of making it go bankrupt. Maintaining this potential satisfaction is of dubious social value. The prospect would not force Westinghouse out of the reactor business, but, it is often asserted by Price-Anderson proponents, it might force out the small sub-contractors. Why, for example, should a valve manufacturer, who can sell his valves elsewhere, risk everything on a reactor accident where the damage might be $1 billion and where his valve might be singled out as *the* weak link? It might be better for him to forgo the reactor business. So runs one of the chief rationales for having a limit on liability. One does not have to be convinced of the validity of this speculative argument to recognize its plausibility.

We do not wish to close this discussion of Price-Anderson without mentioning some of its more positive features. One has already been mentioned—the device of "deferred retrospective" payments to be demanded only after the (improbable) accident occurs. Further advantages lie in the relative ease of payments to claimants. They may receive emergency payments without giving up rights to further claims, and, under a "waiver of defenses" provision, the utilities give up the right to contest claims on grounds of the claimant's negligence.[8]

We also can consider the experience to date with the Price-Anderson insurance scheme. As far as the federal government is concerned, it has been a profit making venture, not a subsidy. In return for the indemnity coverage from the federal government, reactor operators pay up to $90,000 per year.[7] So far, the federal government has collected over $8 million and has paid out nothing.[7] Thus, if one insists on such irrelevant terminology, the utilities have been "subsidizing" the government under Price-Anderson. Of course, should there be an accident and the federal indemnity used, the subsidy would then indeed be reversed. But as the current fees exceed the "fair" premium discussed earlier in the Appendix, the government is getting a bargain even taking this into account.

There is also the private liability insurance pool to consider.[9] Through 1975, the utilities have paid in over $70 million. Payments to claimants have totaled $580,000, primarily in connection with injuries to two workers, neither working at a reactor. The insurance premiums have thus been very much greater than the payments. In consequence, there have been substantial refunds to the utilities, amounting to about 70 percent of the premiums received through 1965 (premiums are held for ten years in a reserve fund, and then partially refunded if the record warrants).

In conclusion, the Price-Anderson Act has worked well to date. The utilities have had to pay only modest insurance premiums, some of which have already been refunded, the insurance companies have gained a large surplus, and the government has gained a small surplus. The public is well protected up to certain limits and stands without formal protection above these limits. It appears very unlikely that these limits will be exceeded. Should they be exceeded, special federal action will be required, as in other cases of national disasters.

We do not hold that the Price-Anderson Act represents a perfect legislative solution to a complex insurance problem. But it appears to be a reasonable attempt in a novel area. The important point is to make the chance of a large nuclear reactor accident very small. Should one nevertheless occur, working out appropriate further indemnification procedures would be a very small part of the problems which would then have to be solved.

REFERENCES

1. *Reactor Safety Study: An Assessment of Accident Risks in U.S. Commercial Nuclear Power Plants,* WASH-1400, U.S. Nuclear Regulatory Commission, (October 1975), Main Report, p. 84.

2. The changes are summarized, for example, in Federal Register, Vol. 41, No. 54 (March 18, 1976).

3. JCAE Report, No. 883, August, 1965 as cited in *AEC Staff Study of the Price-Anderson Act* (January, 1974), p. 26.

4. Ref. 1, Main Report, p. 121.

5. Monthly Energy Review, Federal Energy Administration (November, 1975), p. 27.

6. Ref. 1, Main Report, p. 11.

7. *Questions and Answers on Price-Anderson,* Atomic Industrial Forum Background Information (October, 1975).

8. *AEC Staff Study of the Price-Anderson Act* (January, 1974), p. 11.

9. Testimony of Joseph Marrone, Washington State Legislative Hearings (November 6, 1975).

BIBLIOGRAPHY

The following is a selected bibliography and is by no means comprehensive. It somewhat reflects the authors' preferences, although efforts have been made to include "balance" in the list. We also cannot claim to have carefully examined all books in the list.

Energy—General

Richard Wilson and William J. Jones, *Energy, Ecology and the Environment,* Academic Press (1974).

A Time to Choose, Report by the Energy Policy Project of the Ford Foundation, Ballinger Publishing Co. (1974).

Annual Review of Energy, Volume 1, Jack M. Hollander, ed., Annual Reviews Inc. (1976).

Perspectives on Energy, L. C. Ruedisili and M. W. Firebaugh, ed., Oxford Univ. Press, Inc. (1975).

Energy Perspectives, U.S. Department of the Interior, Supt. of Documents, U.S. Government Printing Office, issued annually.

Energy: Use, Conservation and Supply, Philip H. Abelson, ed., American Association for the Advancement of Science (1974).

Physics and the Energy Problem, American Institute of Physics, Conference Proceedings (1974).

John M. Fowler, *Energy and the Environment,* McGraw-Hill (1975).

Jerry B. Marion, *Energy in Perspective,* Academic Press (1974).

Carol and John Steinhart, *Energy—Sources, Use, and Role in Human Affairs,* Duxbury Press, North Scituate, Mass. (1974).

M. King Hubbert, *Energy Resources,* in *Resources and Man,* W. H. Freeman (1969).

M. King Hubbert, *Man's Conquest of Energy*, in *The Environmental and Ecological Forum*, 1970-71, USAEC, TID 25857.

Nuclear Energy—General

Bernard L. Cohen, *Nuclear Science and Society*. Anchor/Doubleday (1974).

H. A. Bethe, *The Necessity of Fission Power*, Scientific American *234*, 21 (January, 1976).

The Nuclear Power Controversy, Arthur W. Murphy, ed., American Assembly and Prentice-Hall (1976).

David R. Inglis, *Nuclear Energy: Its Physics and Its Social Challenge*, Addison-Wesley (1972).

Ralph E. Lapp, *The Nuclear Controversy*, Fact Systems, Reddy Kilowatt, Inc. (1974).

Nuclear Power and the Environment, American Nuclear Society (1974).

Nuclear Energy—Fuel Cycle, Etc.

Mason Willrich and Theodore B. Taylor, *Nuclear Theft: Risks and Safeguards*, Ballinger Publishing Co. (1974).

T. H. Pigford, *The Nuclear Fuel Cycle*, Annual Review of Nuclear Science p. 515 (1974).

The Nuclear Fuel Cycle, Union of Concerned Scientists, PO Box 289, M.I.T. Branch, Cambridge, Mass. (October, 1973).

John P. Holdren, *Hazards of the Nuclear Fuel Cycle*, Bulletin of the Atomic Scientists, p. 14 (Oct. 1974).

J. G. Speth, A. H. Tamplin, and T. B. Cochran, *Plutonium Recycle: The Fateful Step*, Bulletin of the Atomic Scientists, (Nov. 1974).

Bernard L. Cohen, *Environmental Hazards in Radioactive Waste Disposal* Physics Today *29*, 9 (January, 1976).

Nuclear Energy—Reactor Safety

Ian A. Forbes et al., *Nuclear Reactor Safety—An Evaluation of New Evidence*, Nuclear News, p. 32 (Sept. 1971).

Ian A. Forbes et al., *The Nuclear Debate—A Call to Reason*. Position Paper, Boston, Mass. 1974. Published by California Council for Environmental and Economic Balance, San Francisco, CA.

R. Philip Hammond, *Nuclear Power Risks*, American Scientist *62*, 157 (Mar.-Apr. 1974).

Henry W. Kendall and Sidney Moglewer, *Preliminary Review of the AEC Reactor Safety Study*, Sierra Club and Union of Concerned Scientists (December, 1974).

Report to the APS by the Study Group on Light-Water Reactor Safety, Reviews of Modern Physics *47*, Supplement No. 1 (1975).

Reactor Safety Study—An Assessment of Accident Risks in U.S. Commercial Nuclear Power Plants. WASH-1400, U.S. Nuclear Regulatory Commission (October 1975).

INDEX

Accidents. *See* Reactor Accidents
Accidents, non-nuclear, 115
Actinides, 66
Airplane, nuclear powered, 42
American Physical Society, 86
Atomic Bombs. *See also* nuclear explosions; Manhattan Project, 29; chain-reacting piles, 29; nuclear reactors, 73; explosions, 102; uranium and plutonium, 102; plutonium-240, 102-103; birth of nuclear age, 2
Automobiles, efficiency of, 9
Bethe, Hans A., 24
Black lung disease. *See* Coal
Bombs. *See* Atomic Bombs
Breeder reactors, 93ff, 33, 34, 123; United States program, 33; French program, 33, 94, 124; other foreign, 94; need for, 34; safety of, 95; fast flux test breeder, 94; Fermi breeder, 94; plutonium production by, 95; Doppler effect in, 96-97; efficiency of, 98-99; costs of, 99; liquid metal fast breeder, 97ff; Clinch River breeder, 99
British Medical Research Council, 60, 135
Cesium-137, 38, 66
Chain reactions, 28
Coal, 16; resources, 16; time for exhaustion, 16; U.S. versus world, 18; strip mining of, 28; residues from, 37; uranium in, 37; black lung disease, 145
Cochran, Thomas, 60
Cohen, Bernard L., 61, 70
Commoner, Barry, 14-15
Conservation. *See* Energy Conservation
Consumption of energy. *See* Energy Consumption
Controversy, causes, 51ff, 2
Doppler effect in breeders, 96-97
Einstein, Albert A., 39
Electric power, 45ff. *See also* Nuclear Power; uses, 49; costs of, 45ff; peaking problems, 47-48
Emergency cooling, 139ff. *See also* Reactor Accidents
Energy, 5ff; concept of, 5; units, 5; from sun, 9, 21-25; work as energy, 5; conversion of, 5-6; conversion factors, 5, 129-130; nuclear, *see* Nuclear Energy; food costs, 48; comparative costs, 48
Energy input and output, 52-53

Energy conservation, law of, 8; in the home, 8; costs of, 121, 122
Energy consumption, per capita, 7; human metabolic, 6-7; food system, 7
Energy conversion, thermodynamic law, 8-9
Enrichment, uranium, 29-30
Fission. *See* Nuclear Energy; spontaneous fission, 102-103
Food, in U.S. system, 7; energy for production, 7; costs, 48
Foreign programs, 107
Fossil fuels, 11ff. *See also* Coal, Oil, Natural Gas; use patterns, 11-12; resources, 12; reserves, 12-13; controversy over, 14-15; time for exhaustion, 13ff; costs of, 15; tar sands, 17; oil shale, 17
Frydenburg, Ove, 90
Fuel, reprocessing, 66-67
Fusion. *See* Nuclear Energy
Gandhi, Indira, 103
Geological Survey, U.S., 15
Geothermal energy, 25
Hammond, R. Phillip, 70
Hanford leaks, 70-71
Heat pumps, 24; costs, 48; as a form of solar energy, 24
Hubbert, M. King, 14, 15, 16
Hydroelectric power, 22
Insurance. *See* Nuclear Insurance
International Atomic Energy Agency, 112
Leach, Gerald, 53
Linearity hypothesis, 60, 133ff; applied to safety study, 87, 88
Liquid Metal Fast Breeder. See Breeder Reactors
Manhattan Project. See Atomic Bombs and Nuclear Explosives
Mark, J. Carson, 103, 104
Moderator. *See* Reactors
Mole, Robin, 60
Moratoria. *See* Nuclear Power Moratoria
Natural Gas, residues from, 37; resources of 15-16; usage, 11
Nader, Ralph, 83
National Academy of Sciences, 15
New York Times, 124
Nuclear energy, 27ff; fission, 27ff; fragments of fission, 37-38; delayed neutrons in fission, 38; fission products, 37ff; weight of fission products, 39; energy of fission products, 41; fission products compared with coal and oil, 41; fission of U-235, 41; fission of Pu-239, 41; fusion, 27; status of fusion, 27, 123
Nuclear explosions, detection of, 112; India, 112; safeguards, international, 112
Nuclear insurance, 89, 143ff; Price Anderson Act, 89, 143ff; limits of liability, 144ff; premiums, 143, 147; subsidy to government, 147

Nuclear power, commercial, 42. *See also* Reactors; history of, 42; efficiency of, 43; dispute over, 51ff; net energy of, 52-53; reliability and capacity factors, 53; radiation hazards, 57; need for, 121ff
Nuclear power moratoria, 115ff; domestic effects of, 115ff; international effects of, 118-119; environmental effects of, 116; dangers of, 118-119; California initiative, 115; personnel losses, 118
Nuclear proliferation, 111ff. *See also* Nuclear Weapons
Nuclear safeguards, 101ff; terrorism, 101ff; costs of, 106; bomb fabrication, 102-105; Swedish study, 104-105; transportation, 106
Nuclear weapons, 111ff; *See also* Atomic Bombs
Oil, imports, 1, 13; costs of, 15; world reserves, 15; residues from, 37; consumption, 11, 18; alternative uses, 15; oil shale, 17; tar sands, 17; U.S. resources, 13-15
Ocean, power from, temperature differences, 22
Plutonium, 59ff; toxicity of, 59ff; production by reactors, 59, 95; passage through body, 60; "hot particle" theory, 60ff; ingestion by persons, 60; dispersal by terrorists, 61-62, 105; recycle of, 66; half-life, 66; reactor versus weapons grade, 103; shipments of, 106; Pu-240 contamination, 102-103, 106; Los Alamos exposures, 61; Rocky Flats exposures, 61; experience with, 62; theft, 105
Pollution, thermal, *see* Thermal Pollution; coal, 37, 116, 123; oil, 37; natural gas, 37; nuclear, 37
Price Anderson Act, 143ff. *See also* Nuclear Insurance and Reactor Safety
Radiation, fear of, 2; from nuclear plants, 55, 57; normal environmental, 56; in human bodies, 55, 57; cosmic, 55; units of measurement, 56; in Colorado versus Louisiana, 56; by X-rays, 56; cancer production by, 56-57, 134; linearity hypothesis, 87, 133ff; emission from coal plants, 58; genetic damage by, 89-90; potassium-40, 55-57; from bed partners, 57; from fusion power, 123; BEIR Report, 134-136; NCRP Report, 135-136
Radioactivity, half-life, 65; of nuclear wastes, 65ff; of strontium and cesium, 66; natural, 68
Rasmussen, Norman, 76
Rasmussen Report. *See* Reactor Safety Study
Reactors, 29ff. *See also* Breeder Reactors; light water, 30-33; uranium resources for, 34; high temperature gas cooled, 35; fuel costs, 33; residues from 37-39; residual heat, 38, 73-74; bombs, 73; accidents,

74-76; emergency cooling of, 74, 80-81
Reactor accidents and safety, 73ff, 125; commercial record, 74-76; naval record, 76; Army test reactor, 76; safety study, 76ff; 1957 study, 80; common mode failures, 82; Swedish study, 83-85; pressure vessel failure, 85; Browns Ferry accident, 88-89, 146; insurance for, 89, 143ff; of breeders, 95-98; worst predicted case, 88, 146
Reactor risks, 76ff; for average American, 78; compared to other risks, 78; cancer risks, 87; genetic risks, 89-90
Reactor Safety Study, 76ff. *See also* Reactor Accidents and Safety and Reactor Risks; criticisms of, 85-87; American Physical Society Study, 86-88; Swedish Study, 83-85
Relativity, Theory of, 39
Reserves, 12-13. *See also* individual fuels defined
Resources, 12-13. *See also* individual fuels defined
Safety. *See* Reactor Accidents and Safety
Safeguards. *See* Nuclear Safeguards
Ship propulsion, nuclear, 42
Sierra Club, 86
Solar energy, 21-25; hydroelectric, 22; vegetation, 21; wind, 22; ocean temperatures, 22; director solar (heating and cooling), 22-24; air (heat pumps), 24; generation of electricity by, 24-25
Spontaneous fission, 102-103
Starr, Chauncey, 70
Tamplin, Arthur, 60
Taylor, Theodore, 101, 105
Thermal pollution, 9, 43, 51, 131
Thorium, 35
Union of Concerned Scientists, 86
Uranium, 27-28; enrichment of, 29-30; in granite, 27; in sea water, 28; in Chattanooga shale, 28; on Colorado plateau, 28; isotopes of, 28; U-238 as a fuel, 28-29; conversion to plutonium, 29; resources of, 34; costs of, 33; other uses of, 35; in coal, 37
Wash-1400. *See* Reactor Safety Study
Waste heat. *See also* Thermal Pollution; of nuclear reactors, 43; uses for, 43-44
Waste disposal and waste products, 65ff; strontium and cesium, 66; storage of 76ff; storage on surface, 67; storage in salt deposits, 67, 69-70; storage in pyramids, 70, 125; volume of, 67; hazards of, 68ff; Hanford leaks, 70-71; solidification of, 67, 71; concern for future generations, 65, 71; heat generation of, 67; Pu-240 buildup in, 106
Weinberg, Alvin, 57, 104
Willrich, Mason, 101, 105
Wind power. *See* Solar Energy